CONTENTS

Chapter 1 - Introduction to Cydonia

Chapter 2 - European Space Agency

Chapter 3 - The Regular Tetrahedron

Chapter 4 - The Mounds : Up-close

Chapter 5 - "Cydonia Mound Geometry : A Closer Look" by Professor Stanley McDaniel

Chapter 6 - Previous publications

Chapter 7 - A Walk In The Woods

* * * * * *

APPENDIX A - "The Face"

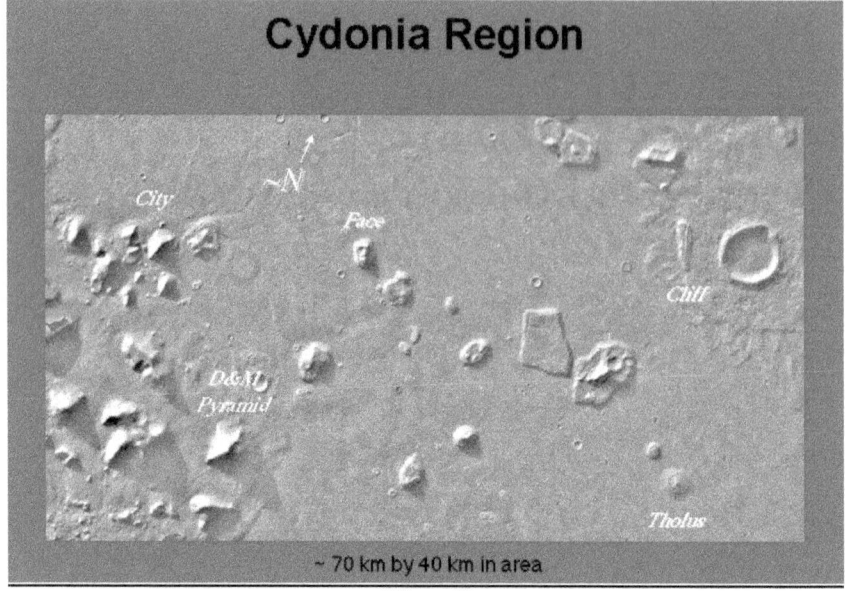

[Figure 1]

[All images courtesy of NASA - Composite enhancement by Dr. Mark Carlotto]

CHAPTER 1 - Introduction to Cydonia

In 1976, NASA, the National Aeronautics and Space Administration, successfully launched two spacecraft towards Mars. Both craft, designated - Viking 1 and Viking 2, were placed into orbit of our neighbouring planet without incident and started to photograph the surface of Mars. Each craft also landed pods onto the Martian soil which conducted several experiments, searching for possible evidence of microbial life. The results of these "life" experiments became very controversial and remain so, to this day in 2020.

The photos of the surface also showed perplexing features and this book is about several of these enigmatic landforms. Is there life on Mars? Was there ever life on Mars, in days gone by when it had a warmer, wetter climate? Did the life there evolve just as life on Earth appears to have evolved and seems to still be in a state of evolution here?

Does Mars have a very advanced civilisation that is currently beyond our detection and comprehension? Or have they already communicated with us via a mathematical, geometric "message in a bottle" across space?

Dedicated to the memory of
late professor of physics Dr. Horace Crater,
who taught physics for 40 years at
the University of Tennessee Space Institute.

===

The author also expresses his gratitude to professor Stanley V. McDaniel, emeritus - philosophy, from Sonoma State University in California who made suggestions for this book, wrote Chapter 5 and is the Founder of SPSR - The Society For Planetary SETI Research. (*SETI = Search for Extraterrestrial Intelligence)

===

Ananda L. Sirisena © 2020

The moral right of Ananda L. Sirisena to be identified as the author of this book has been asserted by him in accordance with the Copyright, Designs and Patents Act 1988.

Other books by this author:

1) MASSIVE VIMANA (UFO) OVER THE ATOMIC WEAPONS ESTABLISHMENT - A Challenge For Parliament

2) SOMETHING STRANGE ON THE LUNAR SURFACE - An Investigation of the crater Paracelsus C

THE MOUND CONFIGURATION IN CYDONIA

AN UNRESOLVED MYSTERY ON MARS

These are interesting questions which can be posed within a scientific framework, or they can be allusions to cultural and religious beliefs, which quite often fall outside of the purview of the scientific method.

The reader is asked to decide: is the mathematics of the "Mound Layout" in Cydonia proof of intelligent life on Mars, or simply a chance, random happening or is it evidence of an ancient civilisation that understood the *inner structure of a three-sided, regular pyramid* - the tetrahedron? Again, the reader must decide whether there is a need to continue investigation into these unusual formations or view them as sheer coincidence. If the reader decides that there is a case for further investigation, then it would be best to bring this information forward to the attention of all space agencies on Earth that are looking to explore the other planets.

The mathematics in this analysis, initially performed by Dr. Horace Crater is simple enough to follow, page by page, as we map the ground layout of the configuration. The late Dr. Horace Crater was a professor of physics at the University of Tennessee Space Institute, where he had taught for 40 years.

Professor emeritus Stanley V. McDaniel from Sonoma State University in California worked together with professor Horace Crater to expand on understanding the layout of the mounds. This book contains a detailed look at the implications, written by professor Stanley McDaniel.

As mathematics is the basis of modern empirical science, then this mathematical analysis done by Crater and McDaniel, is indicative of an *objective* look at a mysterious feature on the surface of Mars and should not be dismissed as *subjective* and therefore not relevant to humanity.

We have just started space travel in a mind-bogglingly huge Galaxy. What will future space explorers find in reality, as opposed to the imaginings of science fiction writers? Are we ready to accept that we have never been alone in this Solar System? Has our science discovered a "message in a cosmic bottle" right next door on a neighbouring planet? Are we ready to accept that as we stand on the shores of a vast ocean of space and peer outwards, we realise that we are indeed tiny specks of cosmic dust and may never have been alone in our Galaxy. And there are millions more galaxies........

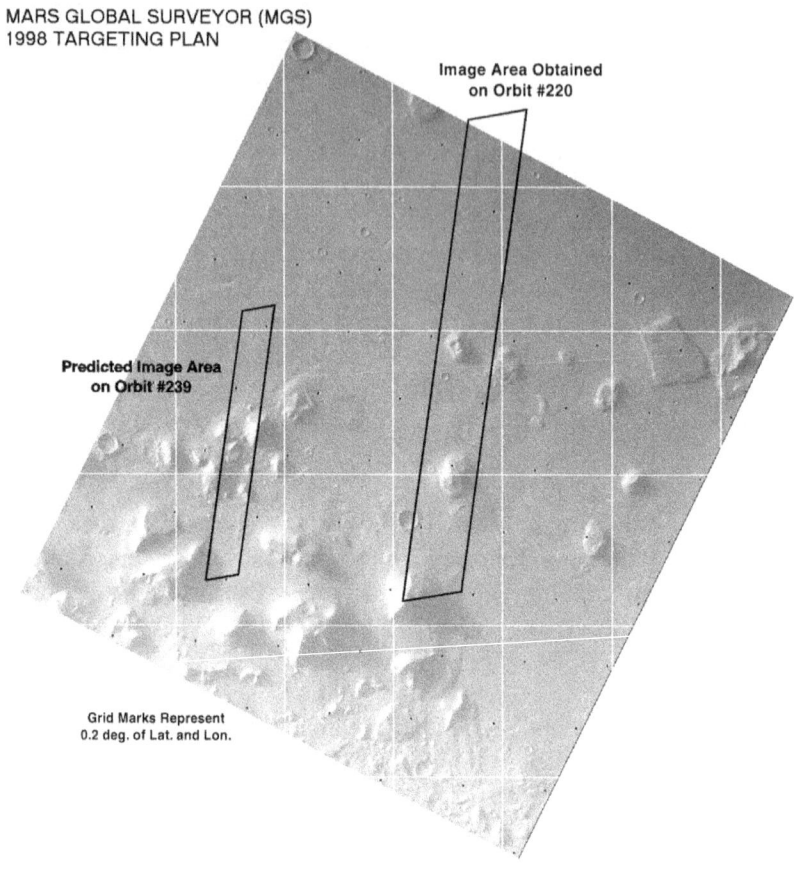

[Figure 2. Plan made in 1998]

The Mars Global Surveyor (MGS) did re-photograph the area twenty-two years later. The MGS mission to Mars confirmed that the features exist in Cydonia and the shiny mounds were evident again in these later images. Naturally there has been a lot of discussion about the composition of the mounds. Some have suggested that they could be eroded pyramids, tors or hills buried in the Martian mud or simply *massifs* in geological terms.

What is of interest here and relevant to Dr. Crater's study is the *placement* of the mounds in a precise mathematical formation, relative to each other. Dr. Crater was aided in his work by emeritus professor of philosophy from Sonoma State University, Stanley V. McDaniel who realised an important aspect of the layout (more about that later).

First, we need to look at the initial analysis and its ramifications.

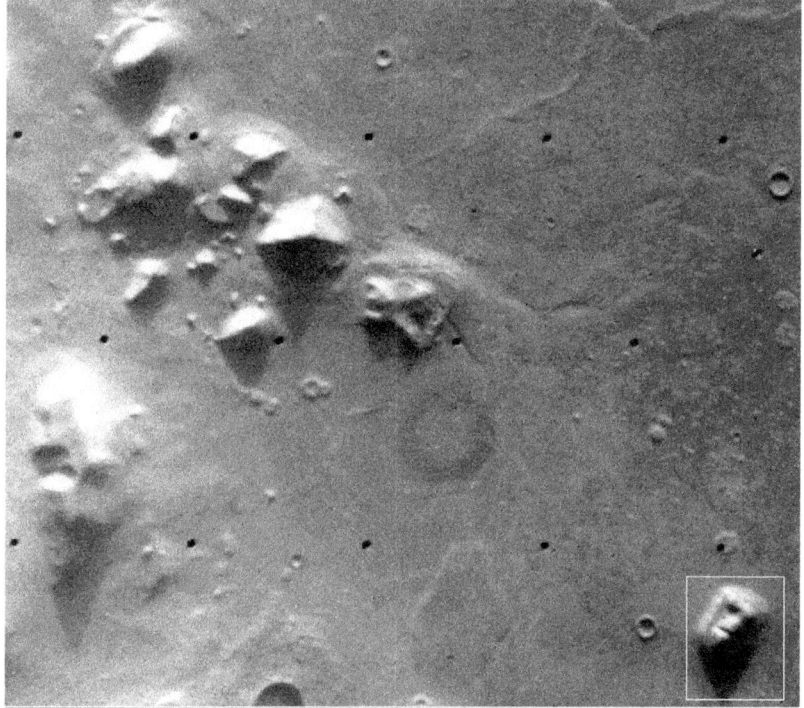

[Figure 3 - Frame 35A72]

[A small portion of one of the Viking 1 pictures - Frame 35A72, orthographically rectified.]

The cropped portion of Frame 35A72 is shown above in Figure 3. This represents an area about 35 miles by 30 miles. The frame numbering was quite simple. This picture was taken on the 35th orbit of the 'A' camera and was the seventy-second in the series, hence 35A72.

The features that are relevant to the present study are the very bright spots, clearly visible in this frame. These "mounds" - as they are designated, are of various shapes, sizes and albedo (which is a measure of reflectivity).
[(Latin: albedo, meaning 'whiteness') is the measure of the diffuse reflection of solar radiation out of the total solar radiation and measured on a scale from 0 (corresponding to a black body that absorbs all incident radiation) to 1 (corresponding to a body that reflects all incident radiation).]

The dark ring in the centre of the frame is not on the surface of Mars but arises from moisture within the lens of the 'A' camera. The black dots in lines across the frame are camera registrations, known as reseau marks.

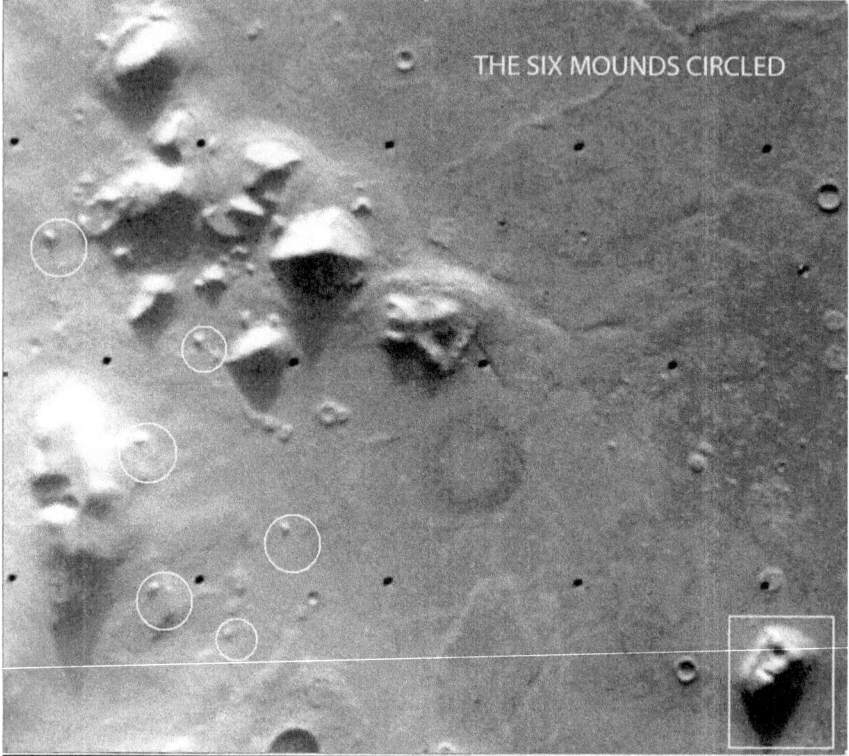

[Figure 4]

Frame 35A72, shown above, has several more of these mysterious mounds than the initial five and six that Dr. Crater used for his careful study. This is basic geometry based on triangles which have a right angle inside them, i.e. an angle of 90 degrees. Right angled triangles are also the core of Pythagoras theorem, which we all learned about in school.

As the reader wades through each diagram, building on the pattern as we move along, it is worth remembering some facts from geometry:

1) All angles of any triangle add up to 180 degrees, irrespective of whether they are right-angled triangles or not.

2) An *isosceles* triangle has two equal angles within it and therefore two equal sides.

3) An equilateral triangle has three equal sides and three equal angles, each of 60 degrees (60 x 3 = 180).

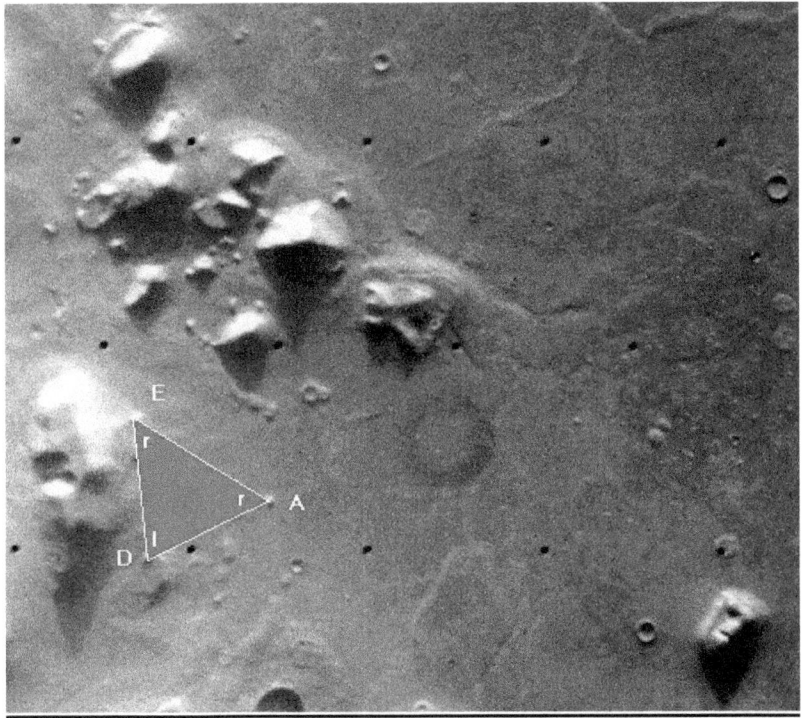

[Figure 5]

Three of the shiny mounds obviously formed a triangle on the surface of Mars. Richard Hoagland had pointed out to Dr. Crater that the triangle appeared to be *isosceles*. That is, two of the sides and angles within this triangle which he named EAD, were equal. This image (35A72) had been orthographically projected.

For the analysis, he termed the angles **l** and **r**. Since, all angles in a triangle add up to 180 degrees, it is clear that:

$$l + r + r = 180$$

Dr. Crater measured angle **l = 70.5 degrees** and **r = 54.75 degrees.**

So, 70.5 + 54.75 + 54.75 = 180 degrees
Another property of an isosceles triangle is that the two sides opposite to the equal angles are also equal. Thus:

$$DE = DA$$

Any three objects must form a triangle. What are the chances that they form an **lrr** triangle? About 1 in 1000, estimated Dr. Crater.

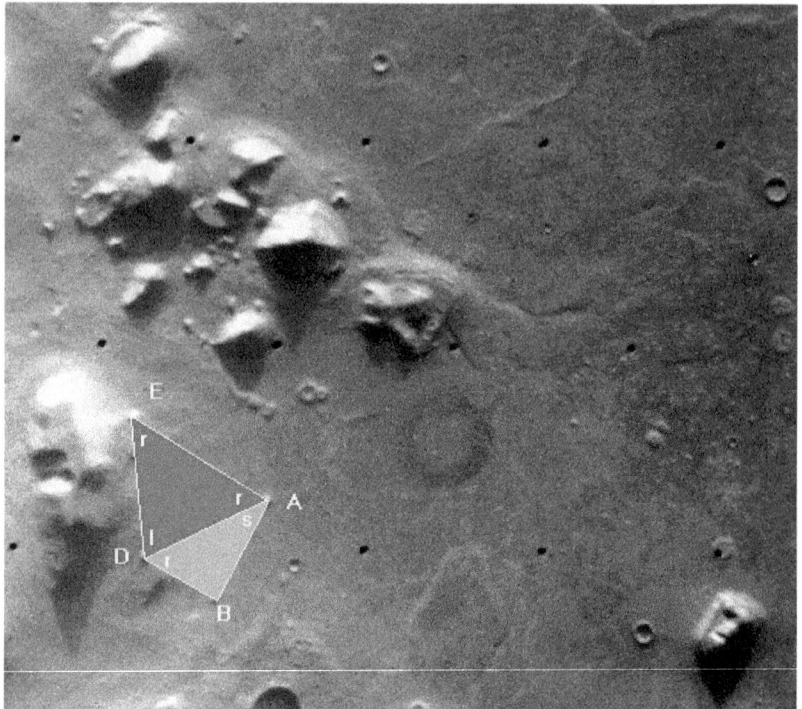

[Figure 6]

Dr. Crater also noticed that when he brought Mound B into his analysis, the **angle r** appeared again. From Figure 6, it is evident that this fact means line EA is parallel to line DB.

This struck him as a little unusual. Also, triangle DBA turned out to be a right-angled triangle. Again, since the sum of all angles in a triangle add up to 180 and angle DBA is 90 degrees, angle **s** was determined to be 35.25 degrees.

So: **54.75 + 90 + 35.25 = 180**

This four-sided shape is essentially triangle EAD placed alongside DBA.

What about all the other shiny mounds? A reasonable question indeed. These mounds appear "shiny" because their tips have a high albedo compared to other features in the general area.

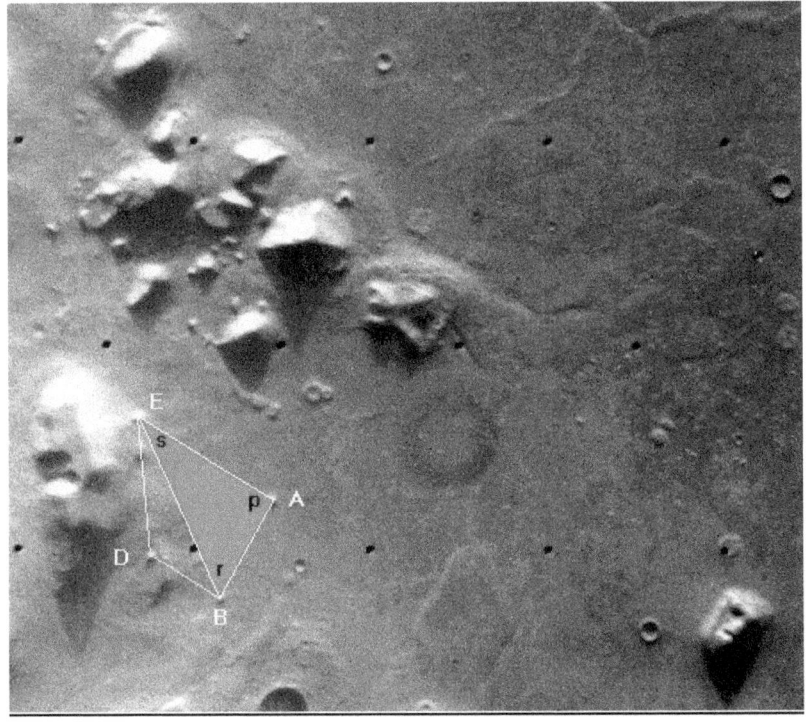

[Figure 7]

Triangle EAB is also a right triangle. Dr. Crater named the right angle of 90 degrees as **p** for his study.

So, **p + s + r = 180 degrees.**

i.e. **(90) + (35.25) + (54.75) = 180**

Dr. Crater's measurements, done on a computer screen, were accurate to the specifications indicated here, i.e. to a quarter of a degree.

The study continued with Horace Crater bringing Mound G into the overall picture. The next image on the next page shows the surprising result.

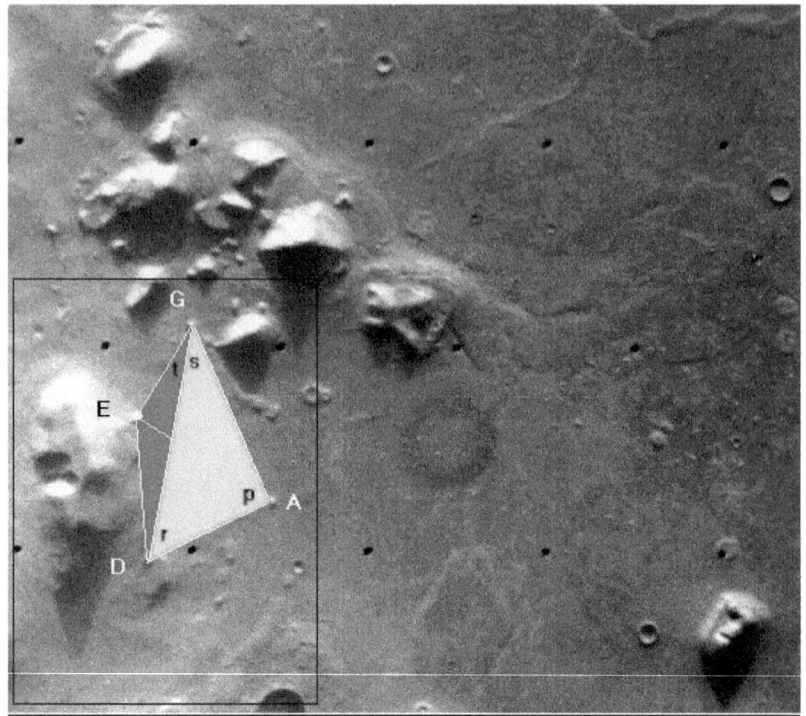

[Figure 8]

As can be seen, from Figure 8 above, triangle GAD is also a right triangle, where angle **p**, as before, is 90 degrees.

So we have another **prs** triangle - similar to triangle EAB (Figure 7).

Let us next look at the formation GADE..

Crater and McDaniel called this formation the "quatrad".

Note that the smaller angle EGD was labelled **'t'**.

The reason why Dr. Crater utilised this naming convention will be made clearer later on.

't' refers to the tetrahedral angle (19.47 degrees or approximately ~ 19.5 degrees)

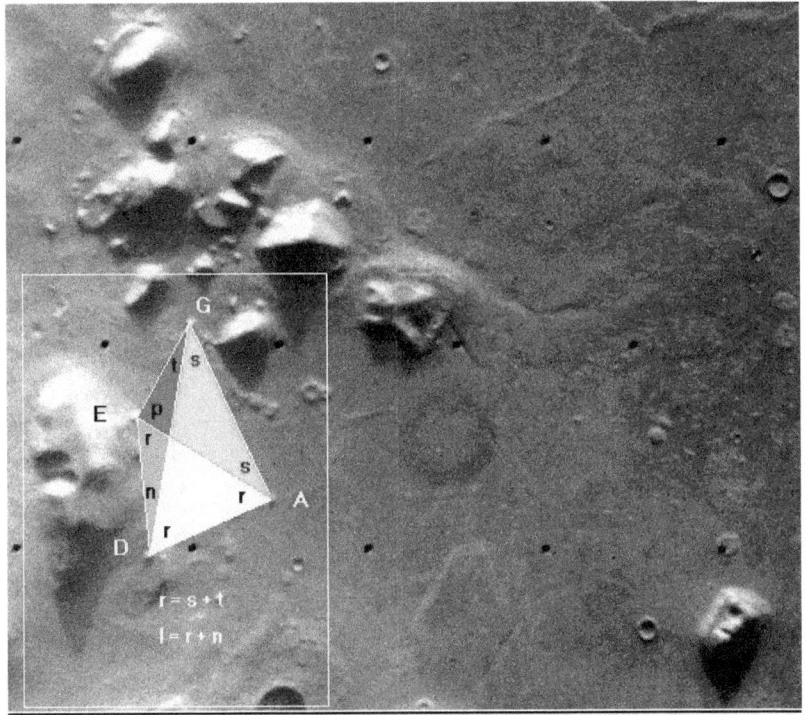

[Figure 9]

The Quatrad GADE.

Let us dissect the quatrad GADE. (Actually, there is another four-sided figure: EDBA. We can look at that later.)

This is composed of triangle GEA abutting the original isosceles triangle EAD.

GEA is another right triangle. We see repeating right triangles and several similar triangles. Similar triangles are defined in geometry as having the same shape, i.e. containing similar angles, even if the triangles are of different sizes.

$r = s + t$ (54.75 = 35.25 + 19.5)

$l = r + n$ (70.5 = 54.75 + 15.75)

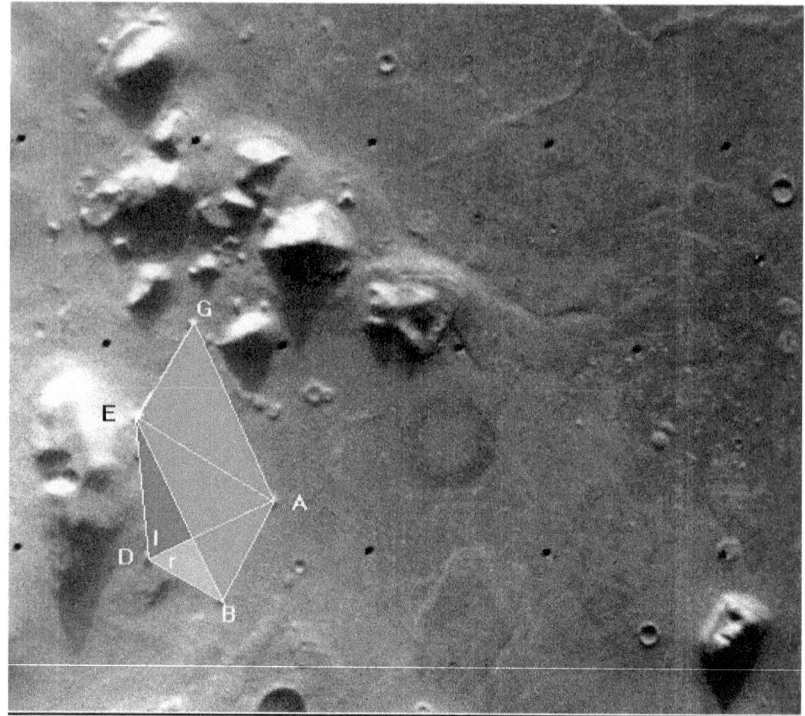

[Figure 10]

Combining Figure 6 and Figure 8 gives the above Figure 10.

Note that GABE forms a parallelogram in this pentad.

This astonishing formation of five mounds: [G - A - B - D - E] was termed **"The Pentad"** by Dr. Crater because of its five-sided shape on the Martian sands.

Note that triangle GEA is similar to triangle EAB. They are identical to each other in the size of the area mapped out by the shiny mounds..

Both angles, GEA and EAB are ninety degrees - termed "right angles" in geometry. Notice that there are two distinct quatrads here: EABD and GADE. If these placements are simply a random arrangement of mounds that are on average, the size in area, of the great pyramid at Giza in Egypt, what are the chances of such "random geology"?

Dr. Crater conducted tests with a random number generator at the University of Tennessee Space Institute (UTSI) to determine the chances of random geology producing such a layout.

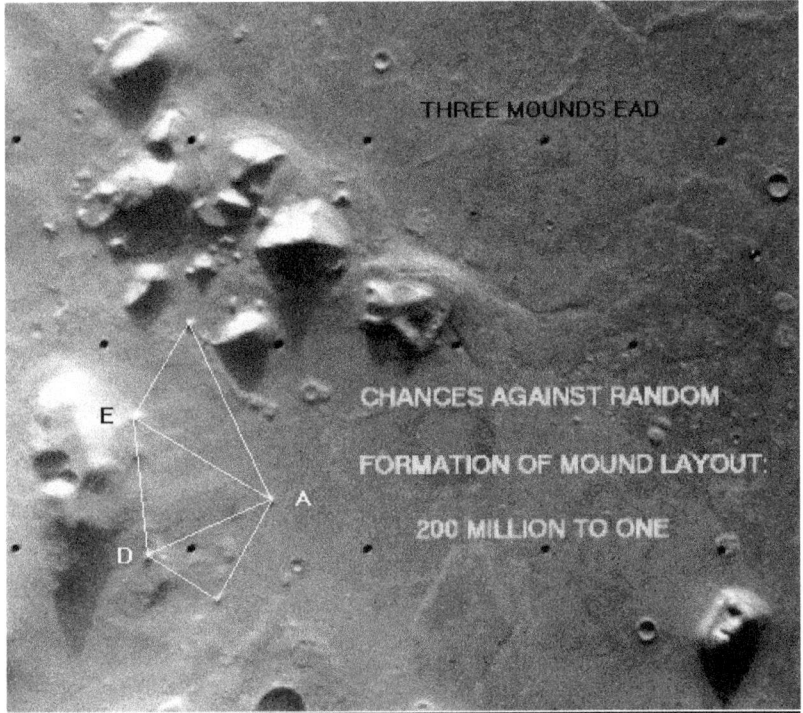

[Figure 11]

Dr. Horace Crater determined the chances against "random geology" forming such a layout as the 'pentad' was 200 million to one. In other words, with the millions of tests done by Dr. Crater using a random number generator, in two hundred million such tests, only one test would result in the particular layout shown above. The precision of the measurements of the distance *between* the mounds was stated by Dr. Crater, to be only a few pixels (picture cells) on the digital photos.

The cameras on board the Viking spacecraft may have been state-of-the-art when they were launched in 1974. Today, they have been superseded by better digital cameras, including those found in millions of cell phones. The Viking photos were printed on paper for study and publicity by NASA. Compared to the resolution achievable with today's cameras, the resolution of the Viking images was a lot less. However, none of Crater's careful measurements have been invalidated by higher resolution photos taken by ESA (the European Space Agency) or later photos taken by the MRO (Mars Reconnaissance Orbiter) launched by NASA. Quite the contrary, they have been confirmed.

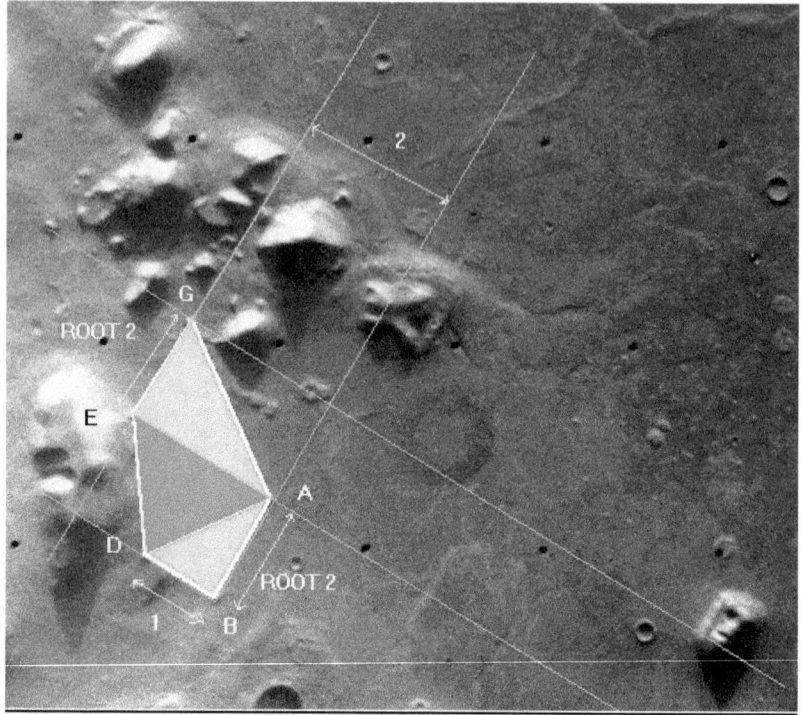

[Figure 12]

The dénouement appeared when professor Stanley McDaniel realised that the mounds sit on a square-root 2 grid. This needs a little explanation. If the distance from mound D to mound B was taken as a unit of 1 (regardless of the actual distance on the surface of Mars in either kilometers or miles), then the distance from mound B to mound A would be the square root of 2.

The square root of 2, written in mathematics as $\sqrt{2}$ or $2^{1/2}$, is the positive algebraic number that, when multiplied by itself, equals the number 2. Technically, it is called the principal square root of 2, to distinguish it from the negative number with the same property. It is calculated to be:

1.41421356237 (or 1.414213562373095) or 1.4142135623730950488...... to a higher precision. The square root of two is an irrational number; it's value cannot be written as the ratio of two integers. So its decimal value as shown above, seemingly goes on, forever and forever....................

Notice that a line through mound G, parallel to line EA cuts across the facial feature.

June 4, 2015. Office of Public Relations,
University of Tennessee Space Institute (UTSI)
news@utsi.edu

Professor Horace Crater Retires From UTSI

Dr. Horace Crater, Professor of Physics at the University of Tennessee Space Institute has retired after 40 years of service. A native of Falls Church, Virginia, he now calls Nashville, Tennessee home. Dr. Crater earned his bachelor degree in physics from the College of William and Mary in 1964 and later earned a Ph.D. in physics from Yale University in 1968.

He was an assistant professor of physics at Vandervilt before joining the Space Institute in September 1975. He was promoted to professor of physics in September 1987. Dr. Crater's field is in theoretical particle physics. He has a deep commitment to the study of very fundamental problems in relativistic quantum mechanics and in quantum field theory, including the relativistic two- and three-body problem as it applies in atomic, nuclear and particle physics. Over the years he has taught a wide variety of advanced courses including quantum mechanics, quantum field theory, solid-state physics and general relativity. He has published numerous refereed papers in the most prestigious physics journals. He enjoys teaching new things as he learns new branches of physics.

A retirement reception was held in Dr. Crater's honor where his wife of almost 52 years, Frances, was in attendance along with faculty, staff, students and guests. Dr. K. C. Reddy, UTSI dean emeritus and professor, read comments by long-time colleague Professor Chris Parriger before saying a few words of his own about Dr. Crater's unique problem solving techniques. UTSI Executive Director Robert Moore then presented Dr. Horace Crater with a service award plaque.

Dr. Crater's plans are to continue teaching courses as needed and to conduct research in addition to more reading and visiting with grandchildren!

June 2014

UTSI Executive Director Robert Moore presents Professor Horace Crater with a plaque for his 40 years of service.

[Figure 13]

In Memoriam

Dr. Horace Crater passed away on 25th October, 2017 at the age of 75. He was born on 26th May, 1942, in Washington D.C. and grew up in Falls Church, Virginia. He went on to attain his Ph.D. in Theoretical Physics at Yale University, graduating second in the class of 1968. He completed two years of postdoctoral work at the Institute for Advanced Study at Princeton University in 1970.

Professor Horace Crater's last scientific paper was published in:

"The Journal of Space Exploration", Volume 5, Issue 3

"The Mounds of Cydonia: Elegant Geology, or Tetrahedral Geometry and Reactions of Pythagoras and Dirac?"

[Figure 14]

Professor Stanley V. McDaniel, author of **"The McDaniel Report"** is also the Founder of The Society for Planetary SETI Research (SPSR).

McDaniel taught philosophy and critical thinking at Sonoma State University in California for over 30 years.

His website is:
http://www.stanmcdaniel.org/

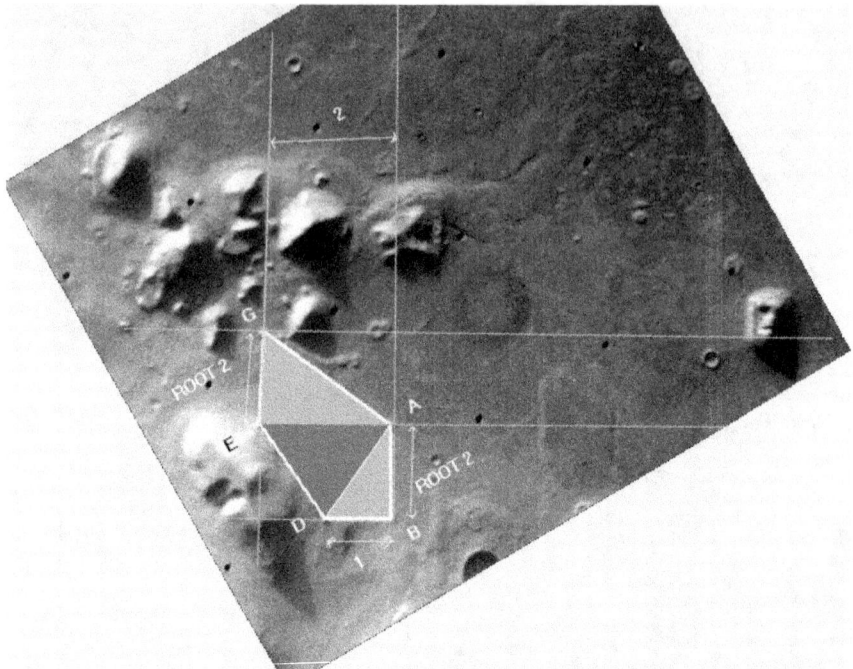

[Figure 15]

The Figure above shows the square-root two grid rotated, so that line BA is pointing due north. The first obvious consequence of this is that a line parallel to EA from Mound G passes just under the chin of the so-called "Face On Mars". Is this a coincidence or is it further evidence of some kind of arrangement here, of cultural import? How ancient are these features on Mars? There have been many guesses, calculations and opinions expressed by planetary scientists but no consensus has been reached about the age of Cydonia. There are experts about the surface of Mars who state that the mounds PGEDBA are "mud mounds" that have remained in pristine condition since they were formed, whenever that may have been.

The cold atmosphere of Mars, with lower air pressure at the surface than that on Earth, somehow has preserved these mounds for many centuries. Weathering from dust storms may have affected them slightly but they retain the position seen, with respect to each other, in later photographs, many decades later.

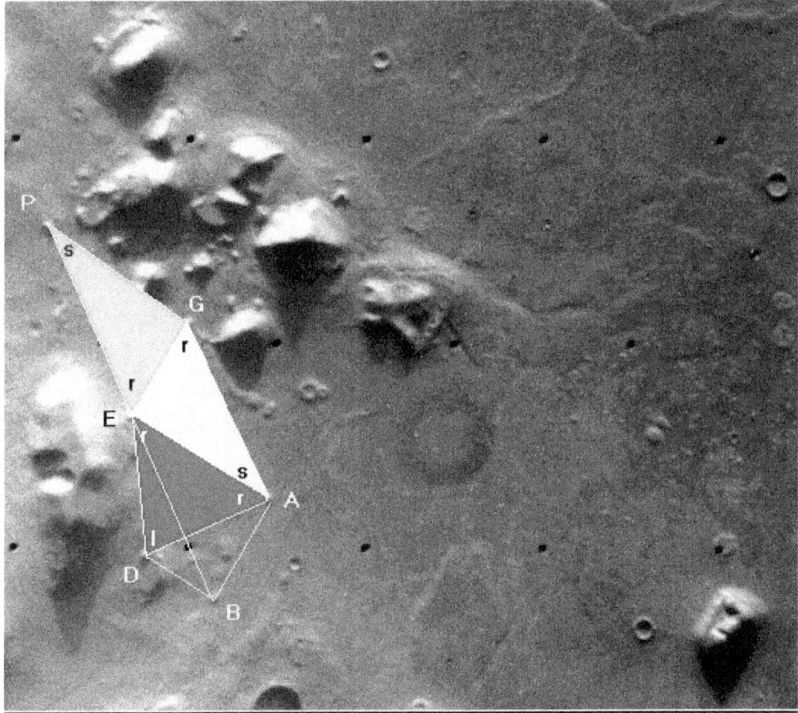

[Figure 16]

The inclusion of Mound P into the analysis done by Dr. Horace Crater, shows something quite extraordinary. In the shaded layout shown above in Figure 16. notice the three similar triangles: PGE, GEA and EAB.

The initial measurements of isosceles triangle EAD has resulted in the above discovery as more nearby mounds were introduced into the study. All research is a matter of studying what is observed. In this case, naturally since the mounds are on a neighbouring planet, one cannot easily send a "ground-truth" team to explore further. We have to depend upon remote sensing probes acquiring high resolution photos over the decades. The fact that the mound configuration has been confirmed by ESA and later by NASA's MRO (Mars Reconnaissance Orbiter) is indicative that Crater and McDaniel were not imagining the layout but were precise in their measurements. To have the results of an experimental analysis confirmed twice is excellent progress.

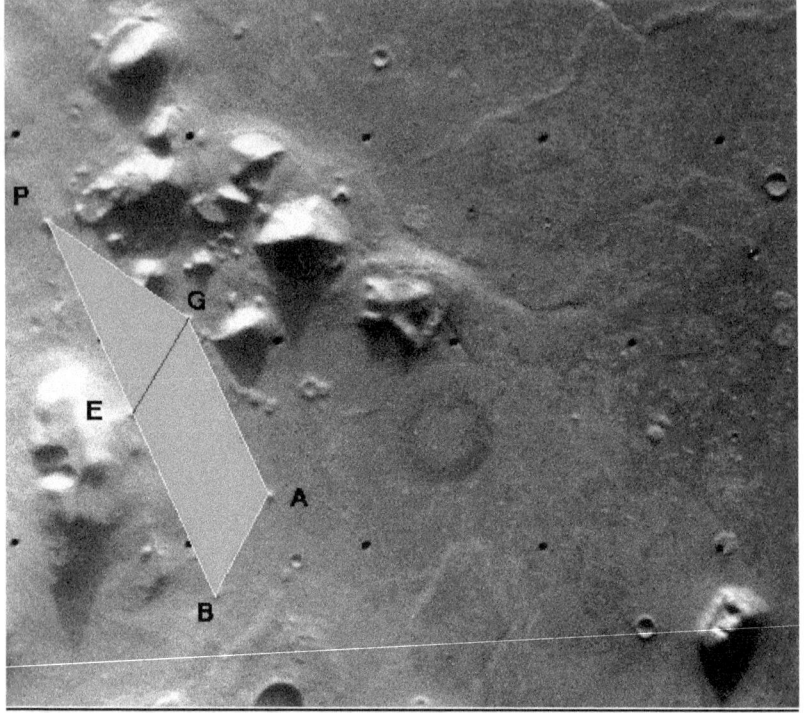

[Figure 17]

When one sees an extraordinary image like the figure above, one cannot fail to be impressed with the visual, visceral reaction that something has been going on in Cydonia. The age of these mounds is not known with any certainty.

The mounds could be millions of years old, regardless of their constitution. They could be mud mounds, they might be the tips of eroded pyramids. The overriding question is: how did they come to form this intricate pattern on the sands of Mars. If such a layout had been found somewhere on Earth, there would have been several exploratory expeditions to the site to excavate each mound and discover their origins. Mars has not had terrestrial men or women on its surface as yet.

There are plans to send astronauts to Mars in the future. Are they likely to visit Cydonia? The colossal expense of a journey to Mars, with at least two spacecraft and four humans in each, for redundancy and aid in case of some unforeseen disaster, might be regarded as unpalatable in the present circumstances of conditions on Earth. It remains to be seen whether we do succeed in sending humans to Mars.

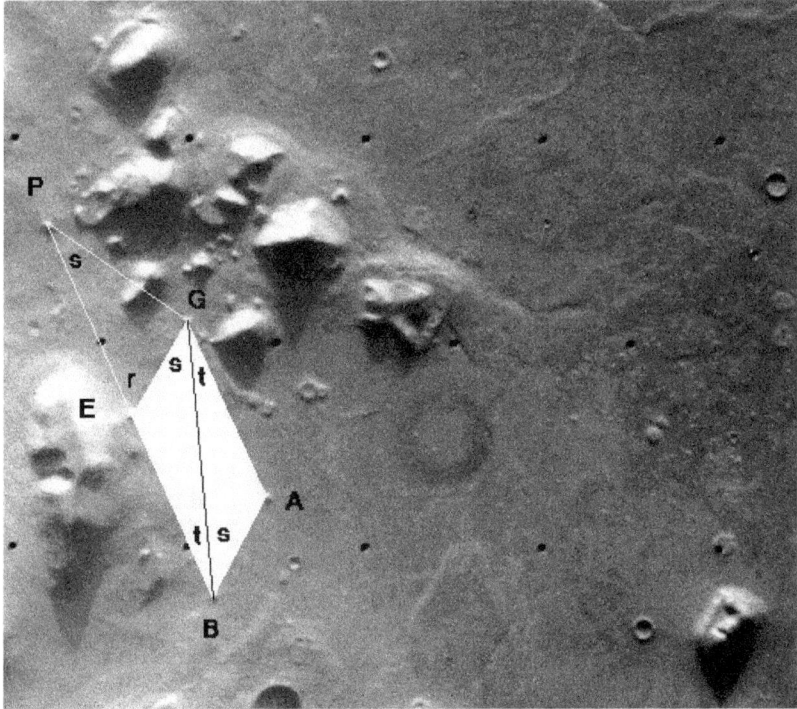

[Figure 18]

From Figure above, we can see that a bisection of parallelogram GABE results in the angles **'s'** and **'t'**.

Recall: **r = s + t,** also derivable from the fact that line EB is parallel to line GA.

Dr. Crater discovered something compelling when he wrote down these angles in degrees as well as radians (180 degrees = pi radians). A circle consists of 2Π radians by definition, so 180 degrees = Π radians.

He found that all these angles could be written in terms of Π and 't'. See the table listing these angles - an implication is that the angle 't' could be some kind of universal constant, such as 'pi' and 'e' and other constants used in physics and mathematics.

The main significance of angle 't' is discussed in the section explaining what happens when a *regular* tetrahedron is circumscribed within a sphere.

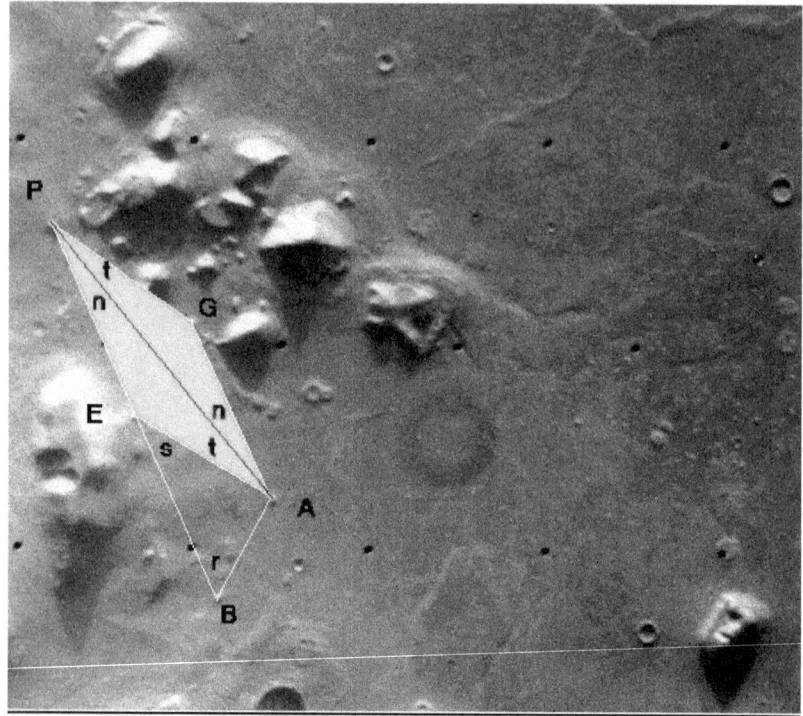

[Figure 19]

Likewise, parallelogram PGAE, as shown above, contains the angles 't' and 'n'. Angle 'n' can also be described in terms of Π and t - see table below:

Angle in degrees	Expressed in terms of *pi* and t
l = 70.5	Π/2 - t
r = 54.75	Π/4 + t/2
s = 35.25	Π/4 - t/2
n = (s - t) = (35.25 - 19.47) = 15.78	Π/4 - 3t/2
p = 90	Π/2
t = arcsin 1/3 ≈ 19.47.......	

The contents of the above table are quite intriguing. The reader who is unfamiliar with this kind of mathematical notation might find the table daunting; however the second column confirms the connections. All the angles in the mound configuration can be expressed in terms of '*pi*' and '*t*'.

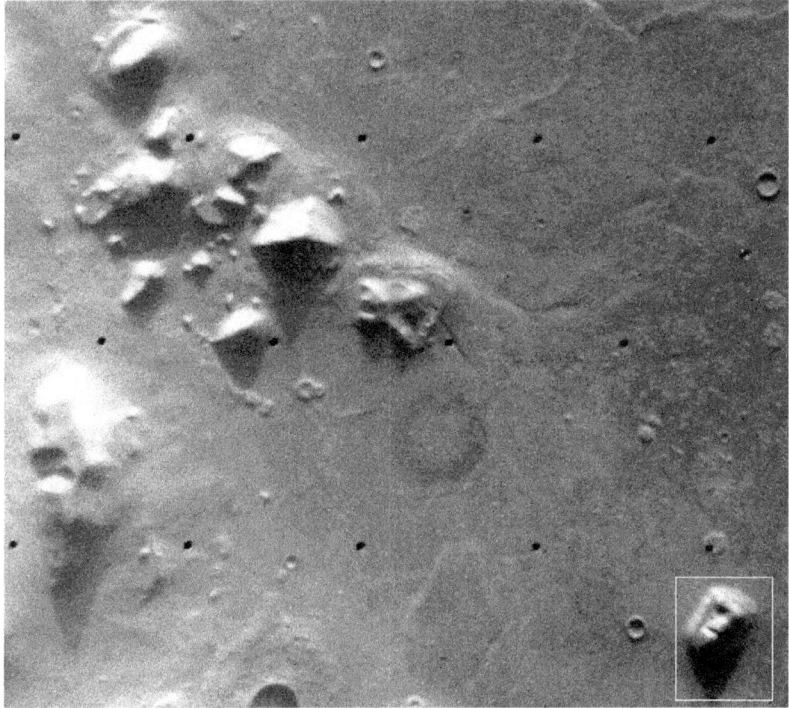

[Figure 20]

The above figure reminds us again of this region of Cydonia, as captured in several images taken by the Viking Orbiters in 1976, without any circling of the shiny features known as "mounds". The mounds used in Crater's study are NOT all of the same shape or size. More detailed images taken by NASA's MRO (Mars Reconnaissance Orbiter) later on show aspects of the mounds in sharp detail and much higher resolution than the Viking Orbiter pictures.

Also note that in the image above, the author has drawn a white rectangle around the famous "Face On Mars". It can be argued that a facial feature is a *subjective* impression, even though all of us everyday recognise faces and do not accuse each other of being subject to pareidolia as we go about our daily lives.

It is noteworthy that the mound configuration in Cydonia is a few kilometres from the "Face", which spawned many books and magazine articles. There is a discussion about the "Face" later in this book.

Dr. Horace Crater [Photo by Ananda Sirisena]

[Figure 21]

On 10th May 2006, Malcolm Smith organised a forum at the British Interplanetary Society. Several papers were presented at this symposium, entitled **"Archeology For Space"**. Malcolm Smith introduced all the speakers and their individual presentations.

After the meeting, professor Crater and myself ventured into a British pub in London for a light drink, whereupon I was able to take this photograph at "The Lord High Admiral" public house.

We had the opportunity to chat about the 'Mound Configuration' in some detail. Crater's brilliant paper delivered at the British Interplanetary Society was titled:
"The Mounds of Cydonia: A Case Study For Planetary SETI".
So SETI, the Search For Extraterrestrial Intelligence is no longer just a search for radio and other signals in the electromagnetic spectrum but a careful search of planetary surfaces through remote-sensed imaging.

[Figure 22]

A photo from the British Interplanetary Society (BIS) symposium in 2006 showing Dr. Mark Carlotto with professor Horace Crater.

[Figure 23]

Shown on left: Chris O'Kane and Thomas Greg Fewer.
Shown on right: Dr. Horace Crater and Dr. Mark Carlotto

List of presentations at the BIS - 10 May 2006

1) **A Sumerian Observation Of The Koefels Impact Event**
by Mark Hempsell and Alan Bond

2) **Archeology's Cold War Windfall - The Corona Programme and Lost Landscapes Of The Near East**
by Keith Challis

3) **Conserving Space Heritage - The Case Of Tranquility Base**
by Thomas Greg Fewer

4) **SETI - Why The Radio Silence?**
by Ananda L. Sirisena, SPSR

5) **Will ET Write Or Radiate? - Interstellar Messages In A Bottle**
by Dr. Christopher Rose

6) **Detecting And Interpreting Patterns On Possible Intelligent Activity In Optical Imagery** by Dr. Mark Carlotto

7) **"The Mounds of Cydonia: A Case Study For Planetary SETI"**
by Professor Horace Crater

l to r: Mark Carlotto, Horace Crater, Ananda Sirisena

[Figure 23 a - Photo taken at The British Interplanetary Society (BIS), May 2006]

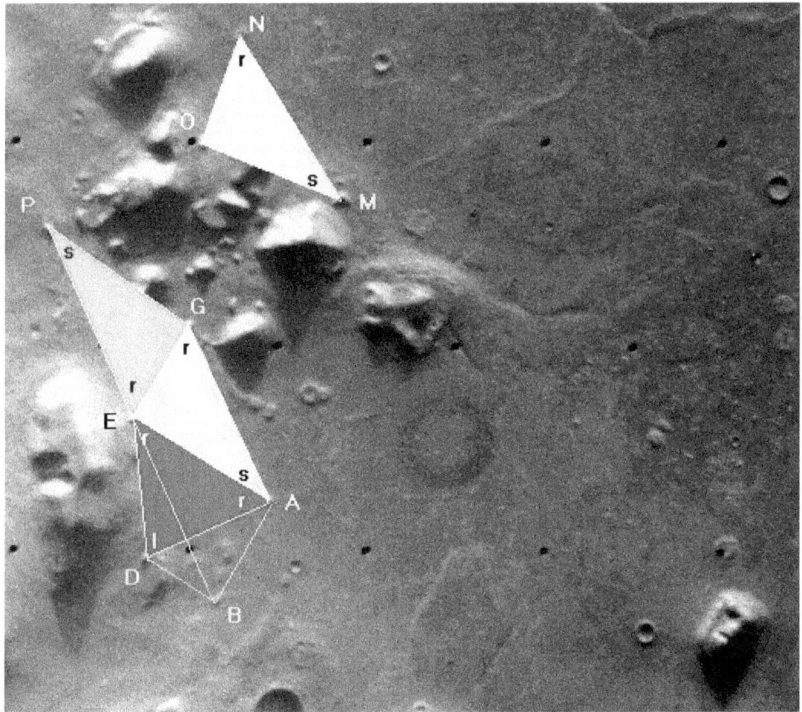

[Figure 24]

In the above 1976 Viking photo, Dr. Crater noticed that there is another right-angled triangle NOM, which is not just similar to triangle GEA but almost identical in size. Both triangles are **prs** triangles.

So triangles PGE, GEA and NOM are congruent. This fact is also very interesting and needs some thought. Notice that there is an angular difference between GEA and NOM, of about 11 degrees. Were these mounds being used to layout astronomical observations, such as at monuments on Earth? Stonehenge comes to mind as well as many other sites throughout the world.

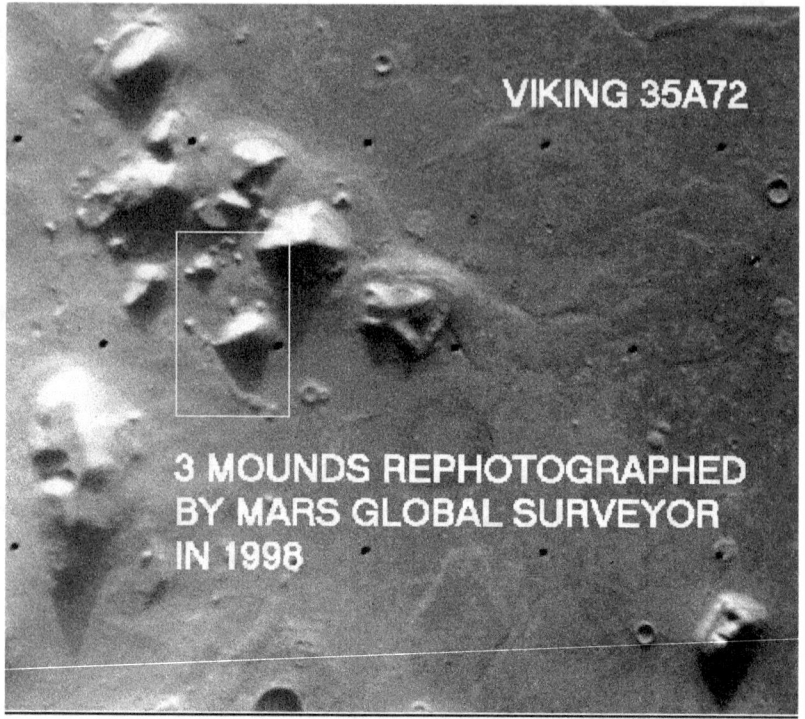

[Figure 25]

Three of the mounds, previously designated as G, J and K were rephotographed in 1998 by the Mars Global Surveyor (MGS). Only mound G forms part of the pentad/hexad arrangement.

Note that Dr. Crater was only studying what was being observed on Mars.

Mounds J and K have been noted to be on lines within the square-root two grid, as shown on Figure 27 and are not part of the pentad/hexad.

The next page shows a greyscale image from the high resolution MGS acquired in 1998.

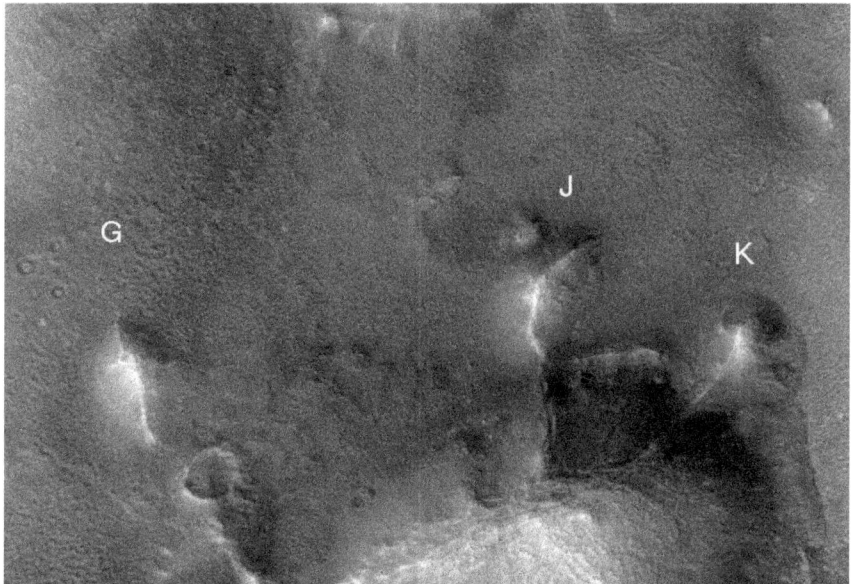

[Figure 26]

The three mounds G, J and K are clearly visible in this high-resolution image released by NASA, from a set taken by the Mars Global Surveyor (MGS) in 1998.

The shiny tops of the mounds stand in stark contrast to the surrounding plains. As stated before, all the mounds are of different shape and size but on average, have a similar square footage.

Dr. Crater used large computer screens to enlarge and observe each mound and its corresponding high albedo pinnacle. If each mound had a square shape it would have been easy to pinpoint each acme. As it happened, this author picked the perceived brightest spot on each mound to enable Dr. Crater to re-perform the fit of the configuration, first seen in 1976, onto the latest images supplied by NASA.

It should be stated here that NASA did not hold back in releasing images that we requested for our study, although they were not released as soon as the images were taken. One wonders - why the delay?

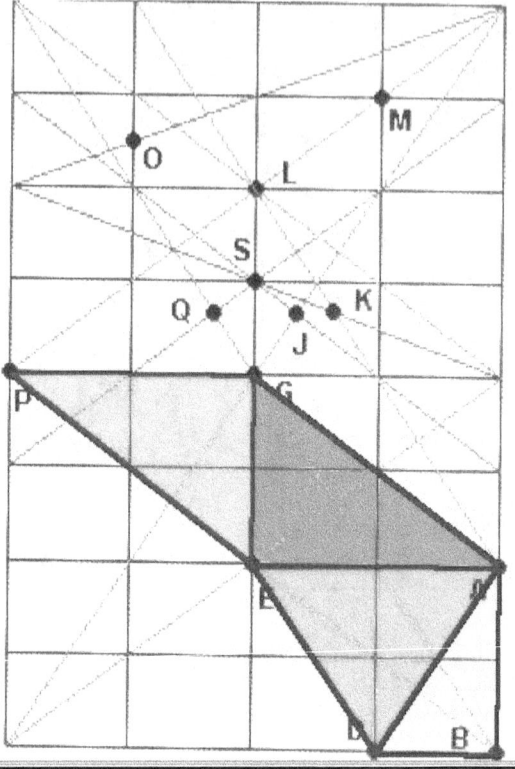

[Figure 27]

The above figure (taken from the book: "The Case For The Face" shows 13 mounds from the region of Cydonia. Part of this diagram was made by Dr. James F. Strange, a professor of Religious Studies at the University of South Florida and formerly a Dean and Chair of the department. James Strange was also the Director of the University of South Florida Archaeological Excavations at Sepphoris, in Israel. In the book, "The Case For The Face", Dr. Strange wrote:

"Several tests applied by Crater to the mound data indicate that the mounds lie in a highly non-random pattern. Following Crater's results, I applied a different test, the Kolgorov-Smirnov test based on intermound distances. My result supported the finding of Crater and McDaniel *that the distribution of the mounds is not random.*"

NB - Mound N falls just outside above representation. of the square-root two grid. I discussed these findings personally with Dr. James Strange in 2008. He passed on in March 2018 in Florida.

Chapter 2

FROM ESA - THE EUROPEAN SPACE AGENCY

===========================

LATER PHOTOS OF CYDONIA ON MARS (2003 - 2006)

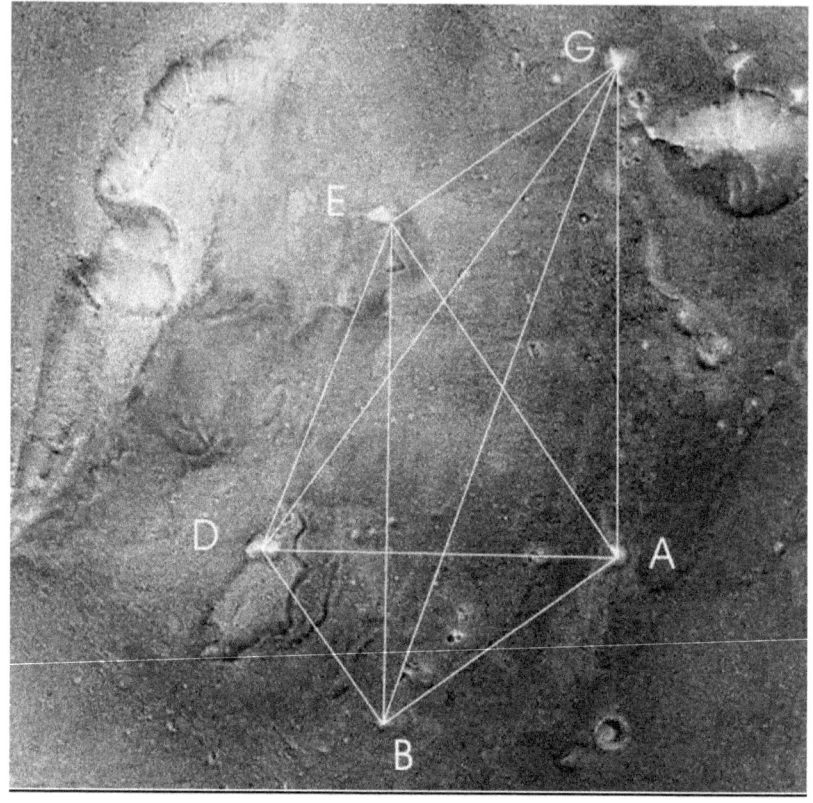

[Figure 28]   [THE PENTAD FROM ESA PHOTOS 2006]

In September 2006, the European Space Agency released a processed colour image of the Cydonia region. The figure above is a portion of this image taken by Mars Express which entered Martian orbit on 25 December 2003.

This image confirmed the discovery made by professors Crater and McDaniel from the 1976 Viking images. Thus, thirty years after the initial NASA images of Cydonia, ESA was able to show, in greater detail that all of the mounds do exist in that northern region of Mars. Their placement with respect to each other is quite apparent in the above composite release from ESA. Albeit, this only shows the pattern from the five mounds, GABDE - the pentad. Later on, we shall be able to view all six mounds once we have expanded on the pentad data, as provided by the European Space Agency.

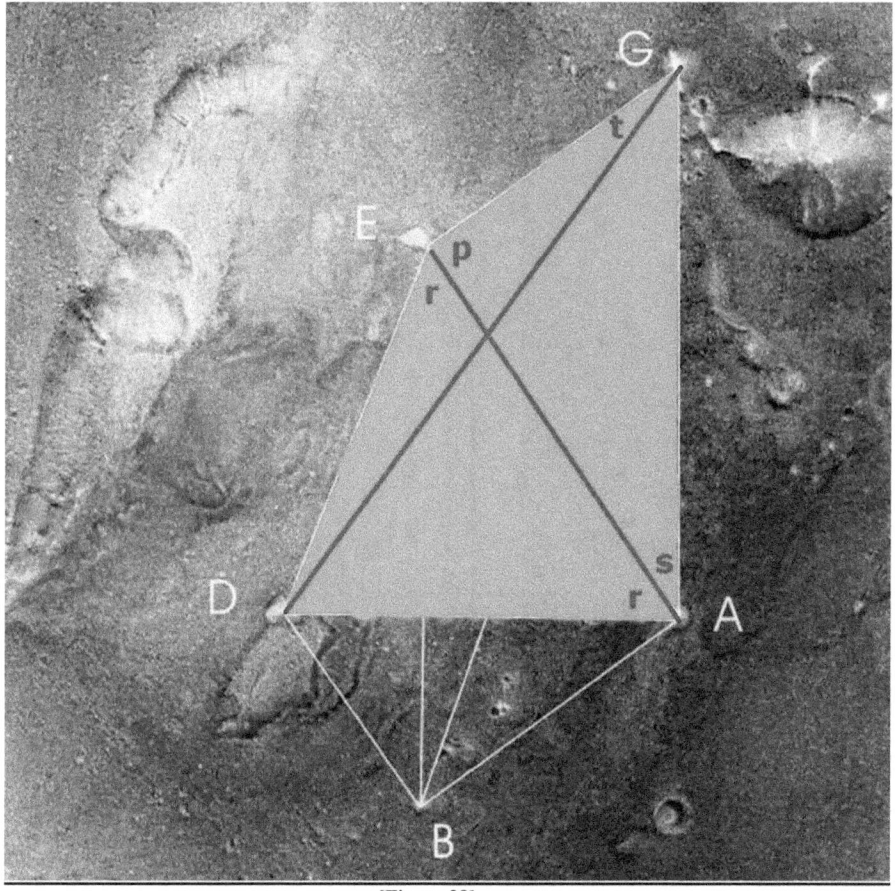

[Figure 29]

Above figure shows the "quatrad" formed by mounds GADE. This four-sided shape consists of the isosceles triangle EAD placed alongside the right-triangle GEA.

One can posit that if an isosceles triangle with the angles **l, r, r,** is then placed alongside a right-triangle GEA, which has angles **p, r, s** - then angle EGD can only be **'t'**.

Although this result could be expressed as **a new theorem in geometry,** the author does not present a rigid mathematical proof for such an assertion. Any mathematician who is interested in providing such a proof is welcome to contact the author by email - (anandals@aol.com).

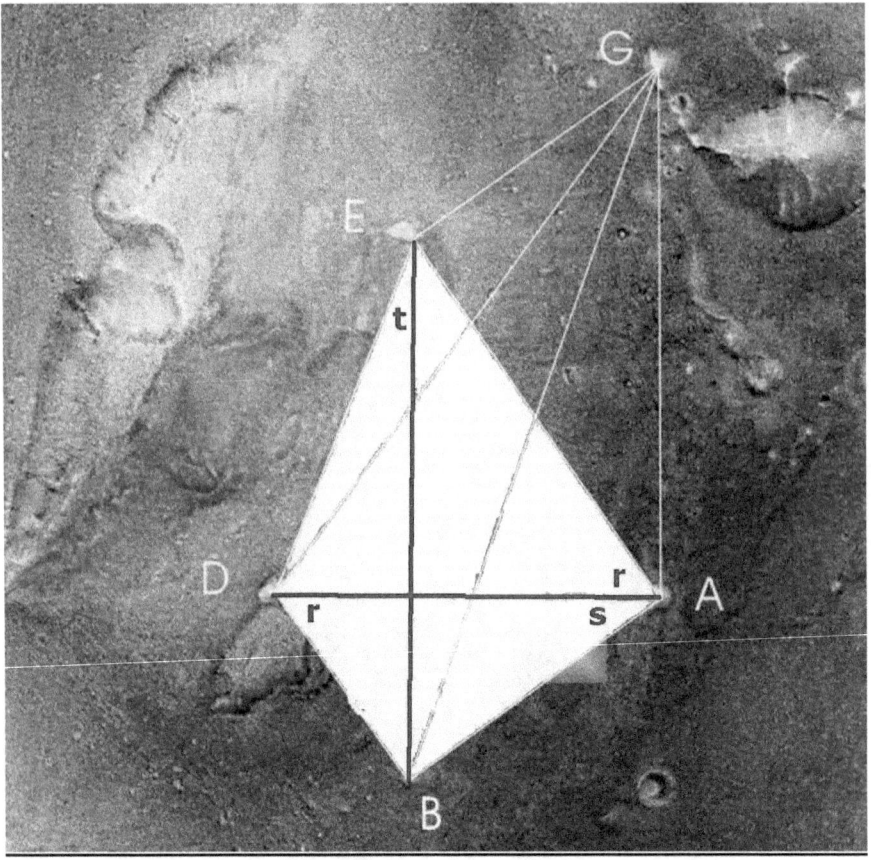

[Figure 30 - Trapezium on Mars]

A similar "theorem" could be posited for the above diagram. If an **l,r,r** triangle EAD is abutted against a right-triangle ABD (whose angles are the same as GEA) and have been carefully measured to be **p, r, s** then angle DEB is **'t'** - and can only be **'t'**.

Note as mentioned before: line EA is parallel to line DB, thus the shape is that of a trapezium, or trapezoid as the shape is known in some books.

What does all this mean? Did some intelligence on Mars lay out these mounds in a deliberate placement for some utilitarian purpose?

Or are we being fooled by some natural geologic process that has resulted in this remarkable pattern?

Dr. Crater calculated that the chances of a pattern such as the pentad GABDE forming naturally is: zero in 200 million tries.

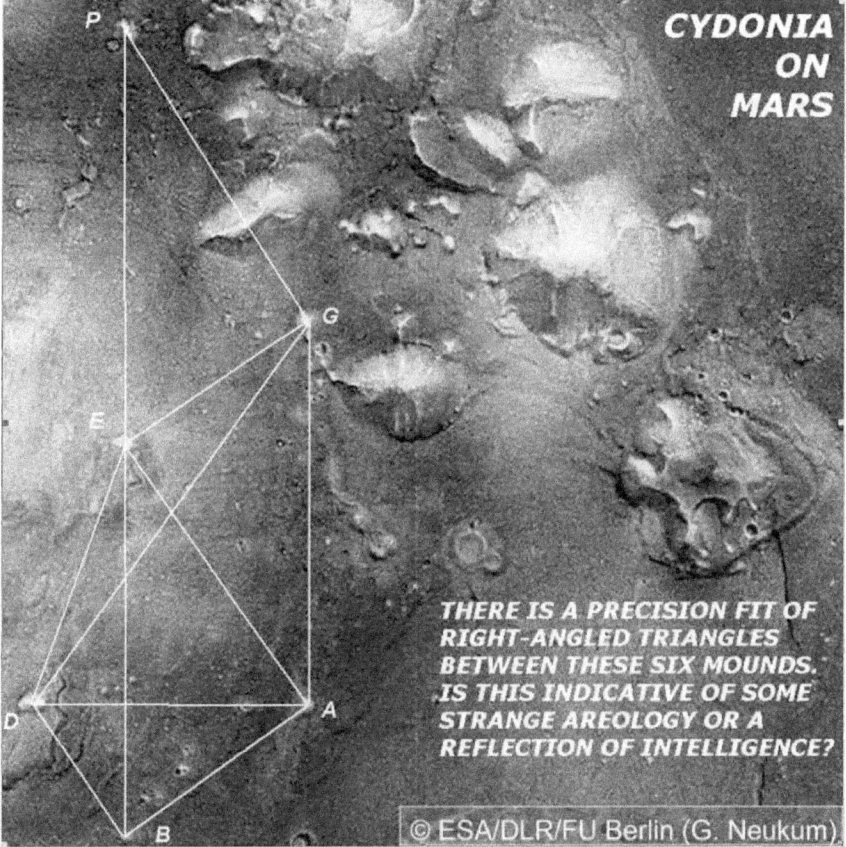

[Figure 31]

An overview of the Cydonia Mensae region as provided by ESA. The hexad is shown here, comprising of the six mounds that played an important part in the Crater and McDaniel analysis. This 2006 confirmation from ESA, of what the NASA Viking pictures showed in 1976, is welcome in further understanding of the possibility of ancient, intelligent life on Mars.

Crater wrote:
"Our computer simulation of the surrounding features and the mound formations themselves demonstrates that the numerous examples of these symmetries, the resultant clustering about certain proportions and the relative precision of the vertices to the mound centres are not compatible with random geological forces."

[Paper by Crater and McDaniel published in the Journal of Scientific Exploration - Volume 13, Number 3, Summer 1999]

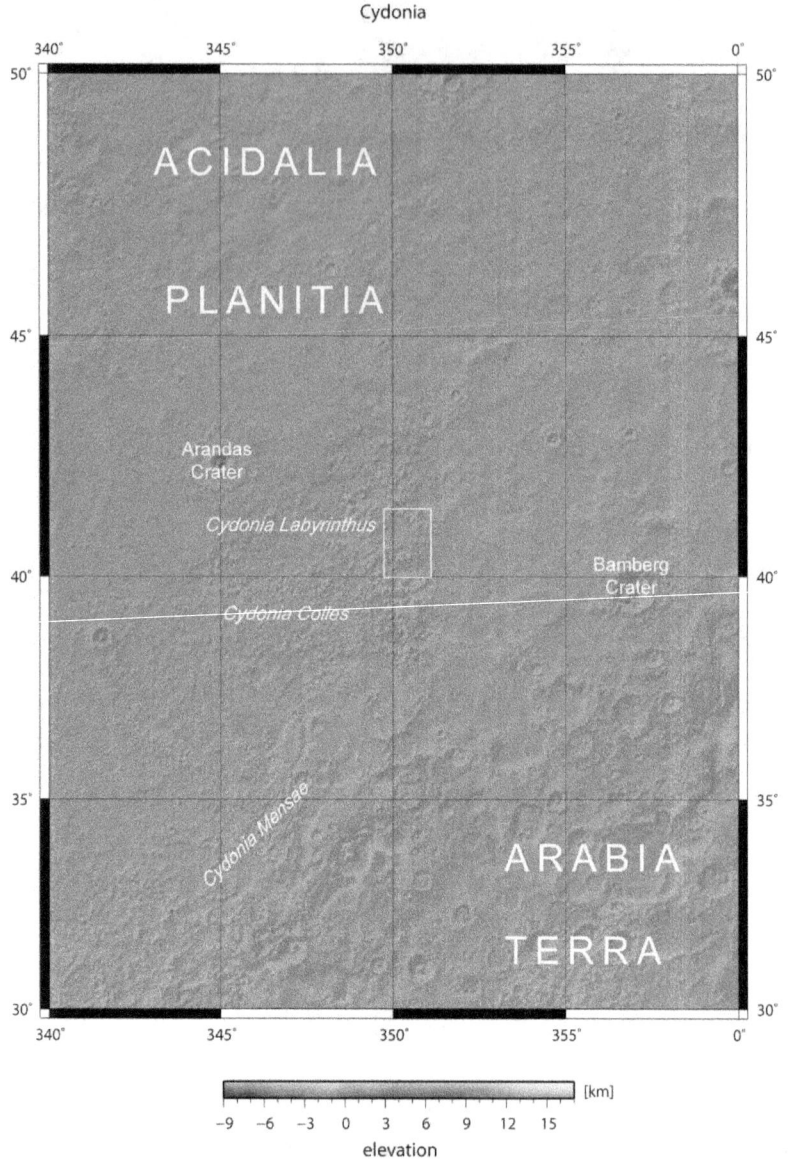

[Figure 32]

The above map, provided by NASA, shows the Cydonia region on the borders of Acidalia Planitia - with Arabia Terra to the south and west and the respective longitudes and latitudes on Mars. Cydonia Mensae may have been along the shores of an ancient lake or ocean.

A 3-D view of the Cydonia region provided by ESA. The six mounds comprising the hexad are shown in this perspective view.

#3d2 Perspektive
[#300 Cydonia (Orbit 3253)]

[Figure 33 - Provided by the European Space Agency]

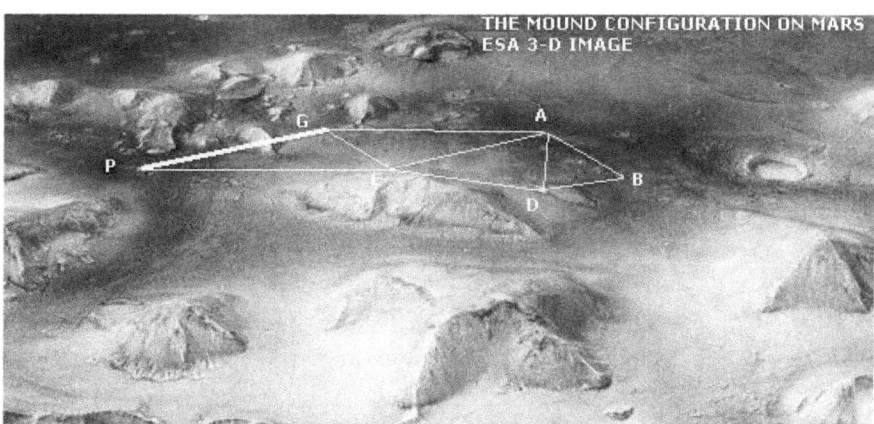

[Figure 34 - Portion of the 3-D image from ESA]

Crater and McDaniel postulated in their paper published in 1999:

"We must conclude that the random geology hypothesis fails by a very large margin, that a radical statistical anomaly exists in the distribution of mound formations in this area of Mars.

"Since previous research in this area seemed to indicate (other) possible anomalies we had reason to focus on this region. If we had chosen an area at random on Mars and found these mound relations then we should factor in the area of the entire planet in our statistical calculations. But this would presuppose that on average all other regions of Mars had a similar density of mounds and that the only mound anomalies are at Cydonia.

"Our studies of numerous Viking images shows that mounds of this type in relatively isolated configurations are far from ubiquitous. The existence of this radical statistical anomaly in the distribution of mound formations in this area of Mars indicates in our opinion a need for continued high-priority targeting of the area for active investigation and determination of the origin and nature of the mounds."

(Journal of Scientific Exploration - 1999)

[Figure 35]

> **MOUND CONFIGURATIONS ON THE MARTIAN CYDONIA PLAIN:**
>
> **A GEOMETRIC AND PROBABILISTIC ANALYSIS**
>
> Horace W. Crater
> Professor of Physics
> The University of Tennessee Space Institute
> Tullahoma, TN 37388
>
> Stanley V. McDaniel
> Professor of Philosophy (Emeritus)
> Sonoma State University
> Rohnert Park, CA 94928
>
> **FOREWORD**
> by
> **ANANDA SIRISENA**
> Copyright 1995

[Figure 36]

The first technical paper about the mound configuration, written by professors Crater and McDaniel was privately published in 1995.

If any reader would like a copy of this historic paper, then please send an email to:
anandals@aol.com
to organise delivery at cost only of printing, postage and packing.

The results published above in 1995 were expanded upon in the Journal of Scientific Exploration in 1999. There have been no valid rebuttals in the Journal.
Professor of Mathemathics, Ralph Greenburg, at the University of Washington, wrote:

"It is worth mentioning an observation of McDaniel. He points out that there could have been a larger collection of mounds whose centers are the points of a grid based on the $\sqrt{2}$-rectangle and which included the twelve mounds studied in the Crater-McDaniel paper. Thus, it is possible that the repetitive appearance of the special triangles in the placement of the 12 mounds is a remnant of a much more highly symmetrical placement of a larger number of mound-like structures."

HORACE W. CRATER is a professor of physics at the University of Tennessee Space Institute, having received his Ph.D. from Yale University in 1968. He is a member of the American Physical Society in The Division of Particles and Fields and the Topical Group "Few Body Systems and Multiparticle Dynamics." His graduate level teaching includes the areas of Classical and Quantum Mechanics, Quantum Field Theory, and General Relativity. His research focuses on relativistic classical mechanics, relativistic quantum mechanics, quantum field theory, and meson spectroscopy. He is the author of numerous peer-reviewed articles on physics in scholarly journals, including "Two-Body Dirac Equations" *(Annals of Physics)*, "Relativistic Naive Quark Model for Spinning Quarks in Mesons" *(Physical Review Letters)*, "Structure of Quantum Mechanical Relativistic Two-Body Interactions for Spinning Particles" *(Foundations of Physics)*.

STANLEY V. McDANIEL is a professor emeritus and former chairman of the Department of Philosophy, Sonoma State University at Rohnert Park, California. He is the Founding Vice-President of the Foundation for Critical Thinking, and a teacher of Logic, Critical Thinking, Philosophy of Science, and Ethics at the university level for over thirty years. His recent book *The McDaniel Report* deals with the ethics and epistemology of research on anomalies in the Cydonia region of Mars. For his classes in logic, he developed an innovative text for introductory Logic students, *Logic and Critical Reasoning*. He is also the author of various research papers and conference presentations in philosophy exploring the relation between philosophy and psychology, including "The Coalescence of Minds" in the volume *Philosophers Look at Science Fiction*. His study guide to *The Philosophy of Nietzsche* has been in print 28 years and was recently released on CD-ROM.

[Figure 37]

The back cover of the first detailed geometric and probabilistic analysis of the subject matter had some biographical notes about the authors.
(Professor Horace Crater passed away in 2017)
Professor McDaniel resides in Santa Rosa, California.

(In some instances this book retains the American spelling of words such as "center" as they appear in US publications but for the English reader the word would be spelt "centre".)

[Figure 38]

This author with professor Stanley McDaniel, shown here in a group photograph taken in San Francisco, California in 1994. To the right of the professor is John Anthony West, to the left is Dan Drasin. The lady just in front of West is the sculptor Kynthia. At the front, kneeling, is Dave Laverty with two other lady presenters at the Whole Life Expo, which was held in San Francisco. The photographer is not known.

[Figure 39 - Stanley McDaniel & Dan Drasin - Photo by Ananda Sirisena]

[Figure 40 - Professor Horace Crater with Dr. Mark Carlotto in Boston - Photo by Mitchell Schwartz]

[Figure 41]

Dr. Mark J. Carlotto with Stephen Corrick

Chapter 3

THE

TETRAHEDRON

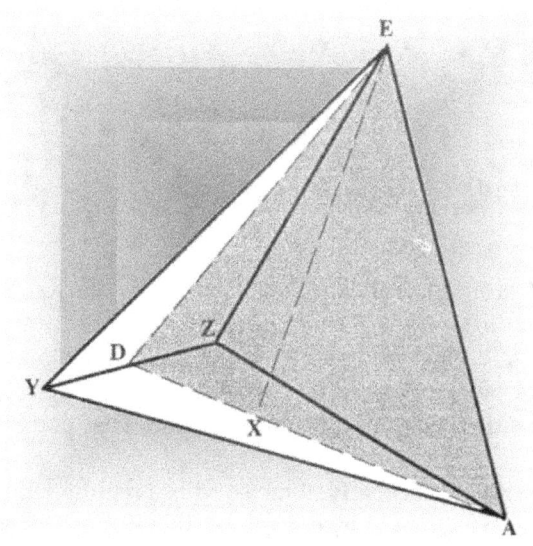

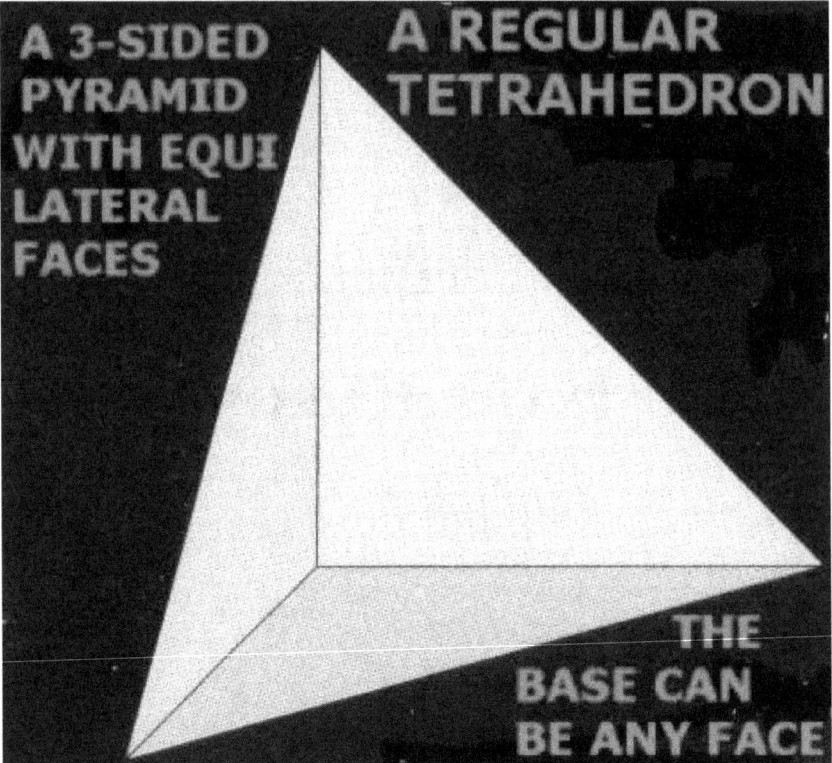

[Figure 42]

A tetrahedron, as a three-sided pyramid, naturally has four faces. A regular tetrahedron would have four equal faces, each composed of an equilateral triangle. Each equilateral triangle would contain 3 equal angles - each of 60 degrees. No matter which way one turns the solid, it will be a three-sided pyramid with an equilateral base and three other equal faces.

In three-dimensional space, a Platonic solid is a regular, convex polyhedron. It is constructed by congruent (identical in shape and size), regular (all angles equal and all sides equal), polygonal faces with the same number of faces meeting at each vertex. Five solids meet these criteria. The five Platonic solids are: The regular tetrahedron (four triangular faces); the cube, a hexahedron (six square faces); the octahedron (eight triangular faces);.the dodecahedron (twelve pentagonal faces) and the icosahedron (twenty triangular faces)

With relation to the mound configuration in Cydonia, on Mars, why are we interested in the regular tetrahedron? Dr. Crater was aware that if one bisects the first Platonic solid, one gets a triangular face with the same angles as the triangle EAD, namely **l, r, r.**

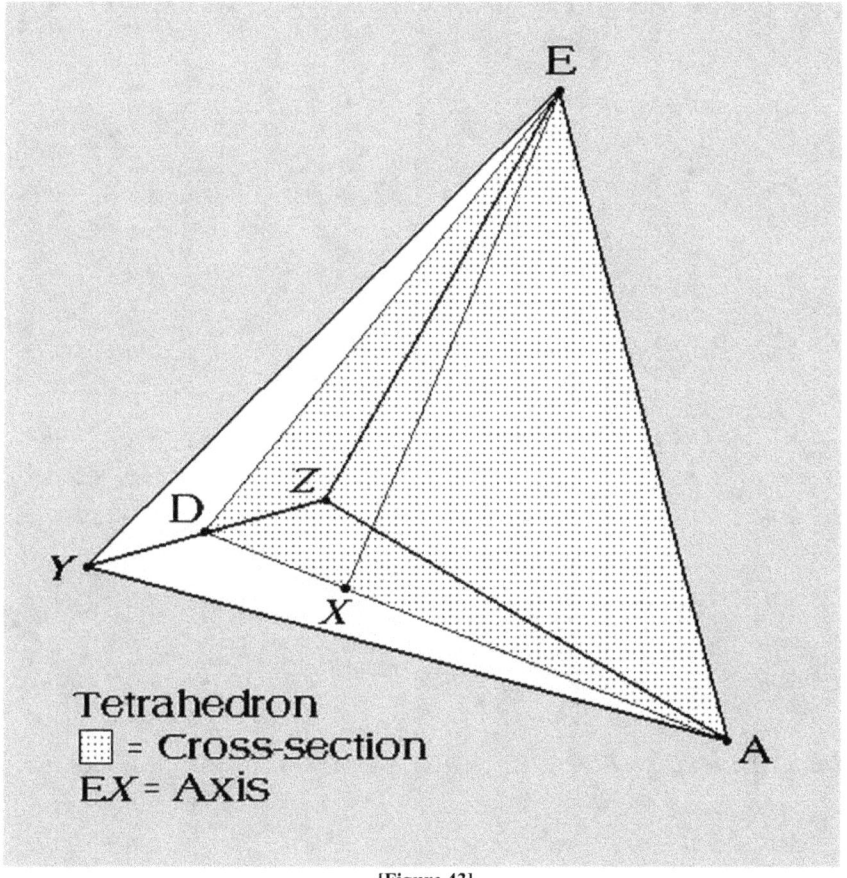

[Figure 43]

In the above sketch, the regular three-sided pyramid is cut in half along the lines DE and DA.
It seems that whoever laid out the complex at Cydonia was aware of the internal geometry of a regular tetrahedron.

If one polls the human population, it will be discovered that not everyone on Earth knows this simple fact. Only a small proportion of humans are knowledgeable about the internal structure of a regular tetrahedron. This is something that I have discovered during lecturing about the subject throughout the world.

Very few people know the details about the inner structure of a three-sided pyramid.

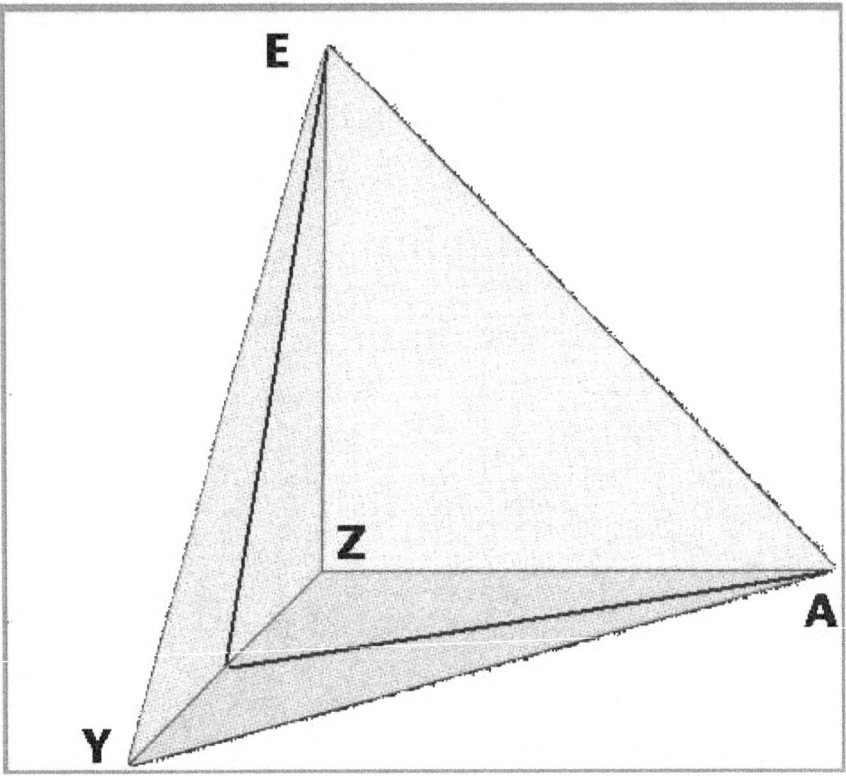

[Figure 44]

In order to fully understand the tetrahedral angle **'t'**, consider the above diagram. The letters E, A, Y and Z represent the four vertices of the regular tetrahedron. Such a tetrahedron contains 4 vertices, 6 edges (EY, EZ, EA, AZ, YZ and AY) and 4 faces. The 4 faces each consist of equilateral triangles, EAZ, EYZ, AZY and EAY.

So, EYZ, EAZ, AZY are the three visible faces in our diagram. The fourth face EAY is not visible to us in this 3-dimensional representation. Now, let us bisect the edge Y - Z. The midpoint of that line can be marked as D, as shown in the next diagram. If we bisect this solid shape at D, we will have cut the tetrahedron into two parts - each with the face EAD.

Dr. Crater shows, in his technical paper, that the angles in this bisection EAD are identical to the angles he measured in his study, starting with triangle EAD, namely **l, r and r.** Is this a coincidence?

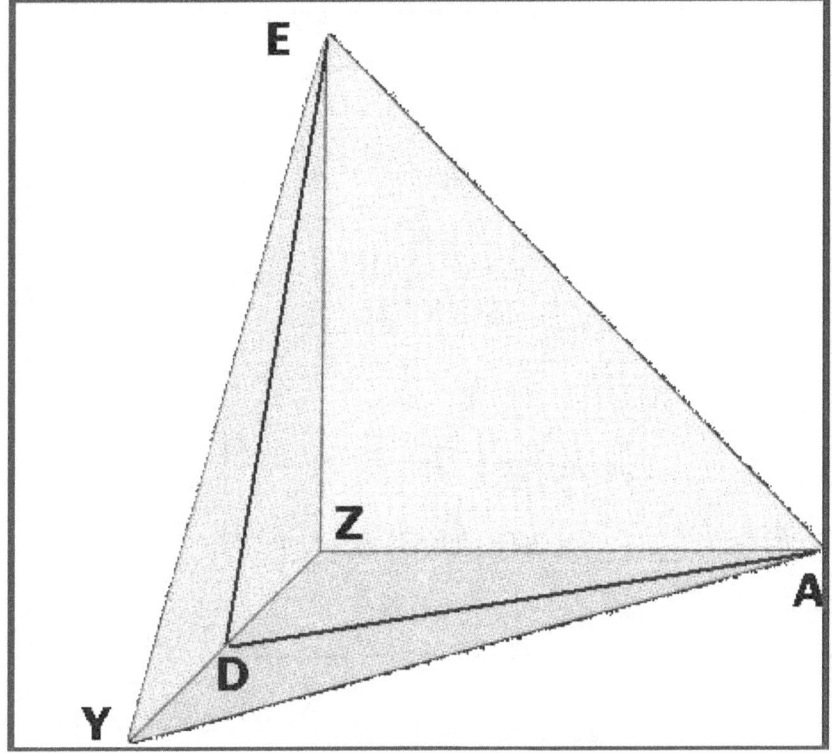

[Figure 45]

Perhaps it is a coincidence but we do not know for certain. Future explorers of Mars, who are likely to land in Cydonia, will no doubt investigate the true structure of each mound in order to ascertain how they ended up displaying such precise locations, so as to make the ground layout visible in our remote sensing photographs.

Crater and McDaniel, prove the following fact mathematically:

If a regular tetrahedron is encased in a sphere, such that one vertex of the tetrahedron is at the north pole - top of the sphere, the other three vertices will touch the sphere at latitudes just below the equator of the sphere, actually at 19.47 degrees below the equator.

Angle **'t'**, 19.47 degrees is thus given the nomenclature ***"tetrahedral angle"***. This fact is not apparent. Intuitively, it may seem that the other three vertices would touch the sphere at its equator but that is not the case. Truly fascinating. The geometric message in the mound configuration at Cydonia appears much deeper than first thought.

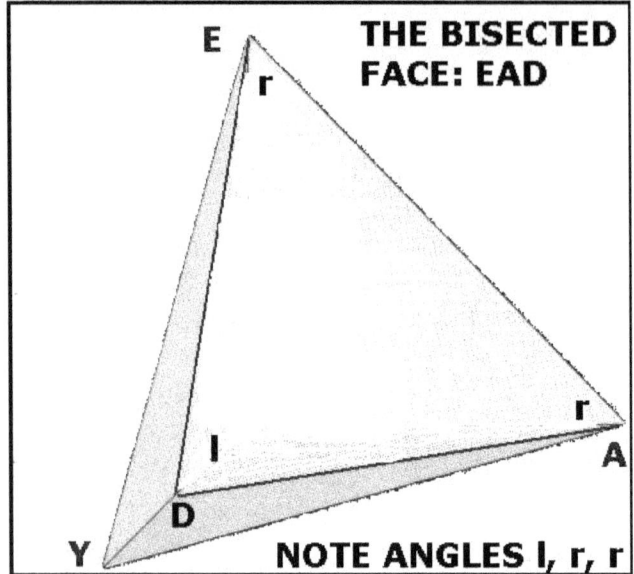

[Figure 46]

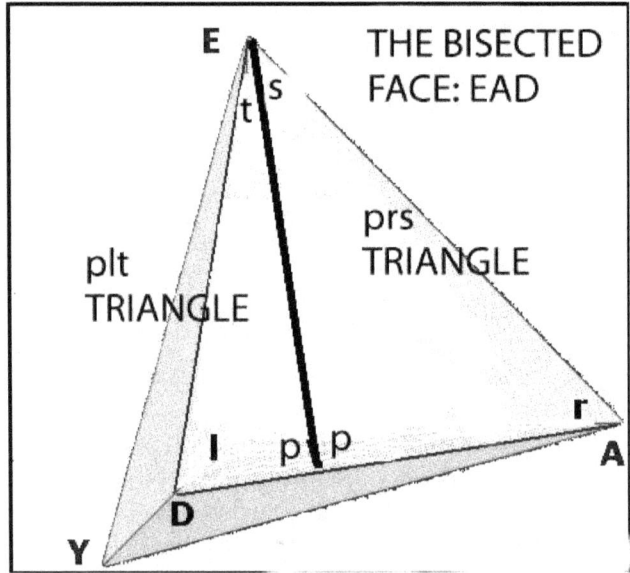

[Figure 47]

Dropping a perpendicular from apex E to edge DA, on the face EAD results in a p-l-t triangle alongside a p-r-s triangle. We have seen all of these angles before; NB: p = 90 degrees

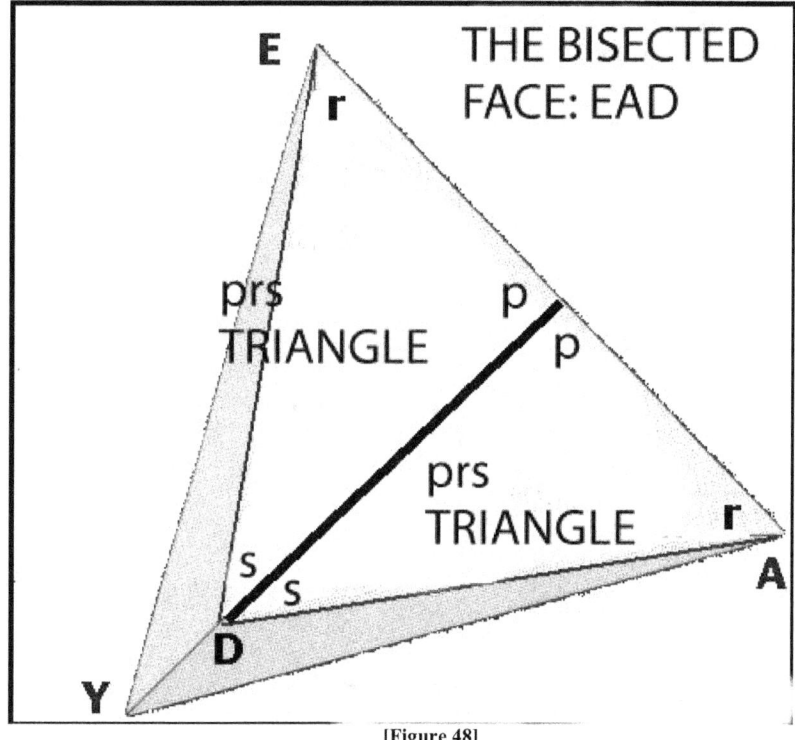

[Figure 48]

Dropping a perpendicular from D to edge EA results in two **p-r-s** triangles where **p** is the right angle: - 90 degrees.

Notice what happens if one drops a line from D to EA. We end up with two congruent **p-r-s** triangles, where p=90 degrees, as before.

We can see that the triangles discovered by Dr. Crater in the mound layout appear in the innards of the sliced tetrahedron triangle EAD. Is this sheer coincidence or the result of intelligent layout. No doubt, the arguments and scientific debates about this perceived pattern will continue on into the future years and decades.

Future explorers on the Martian surface will add to our knowledge of the science of "shape power".

THE METHANE MOLECULE

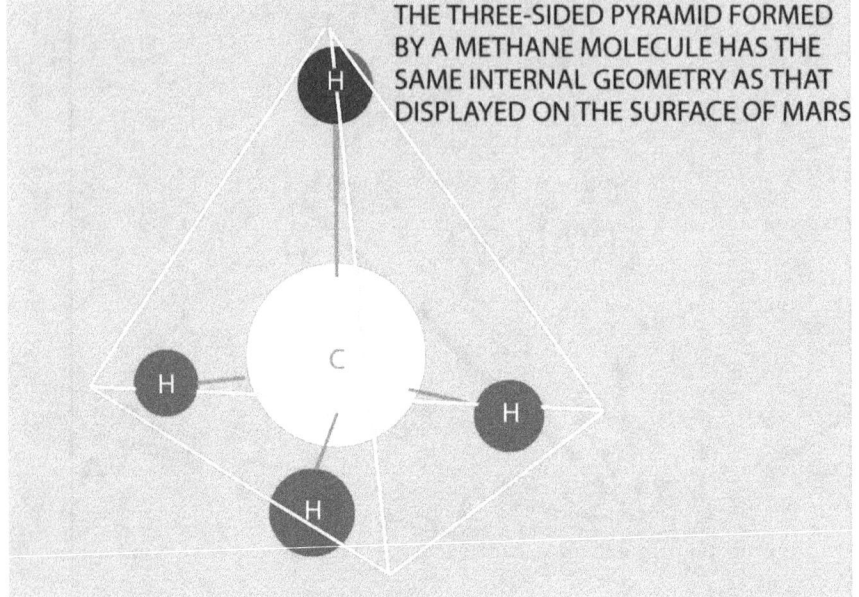

[Figure 49]

The regular tetrahedron is also of some import in the subject of chemistry, for it is the shape of a methane molecule (CH4). One central carbon (C) atom bonds with four hydrogen (H) atoms. Chemists are interested in the angle formed by two hydrogen atoms in relation to the central carbon atom.

This angle has been derived to be 109.47 degrees, approximately. Readers who have followed the page by page display of the mound configuration so far, will recognise that in degrees, 90 + 19.47 = 109.47 (**p + t = 109.47**), using Dr. Crater's terminology.

The angle is generally shown as **109.5** degrees in chemistry textbooks, as an approximation, to one decimal point..

Carbon-based life molecules are also made from carbon atoms bonding with other atoms.

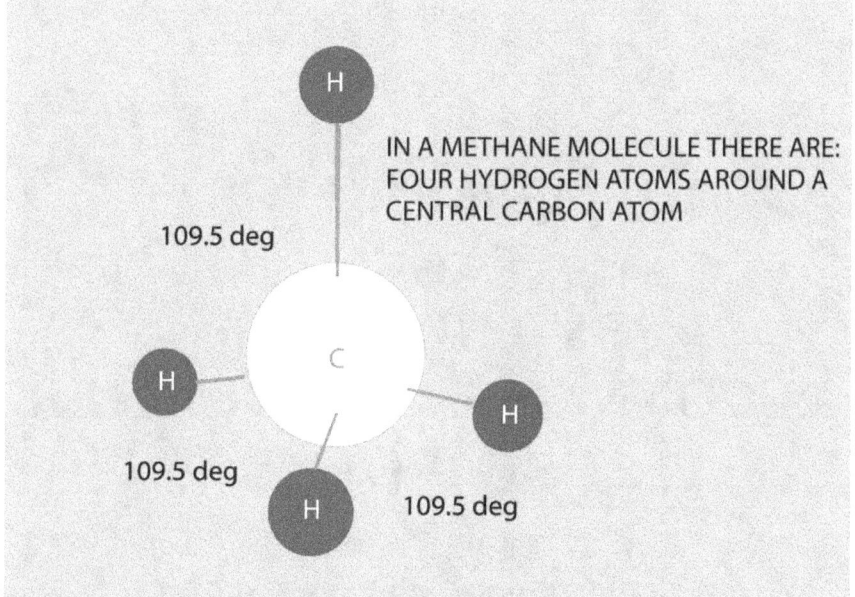

[Figure 50]

All bond angles, between the carbon atom and each hydrogen atom, are 109.5 degrees,
>	i.e. 90 + 19.5.

This reinforces the fact that if a regular tetrahedron is encapsulated by a sphere, such that one vertex of the pyramid is at the north pole of the sphere, then the other three vertices touch the sphere at a latitude just below the equator - at 19.5 degrees. This fact emerges by virtue of the internal geometry of such a three-sided pyramid and its encircling sphere.

The three hydrogen atoms in the lattice lie around the carbon atom at 19.47...degrees below the plane of the carbon atom.

Chapter 4

LATER PHOTOS TAKEN BY THE MARS RECONNAISSANCE ORBITER (MRO)

(PHOTOS COURTESY : NASA/JPL/ASU/MSSS)

MOUNDS

A, B, D, E, G and P

IN HIGHER

RESOLUTION

[[Figure 51 - SHOWING MOUNDS A, B AND D]]

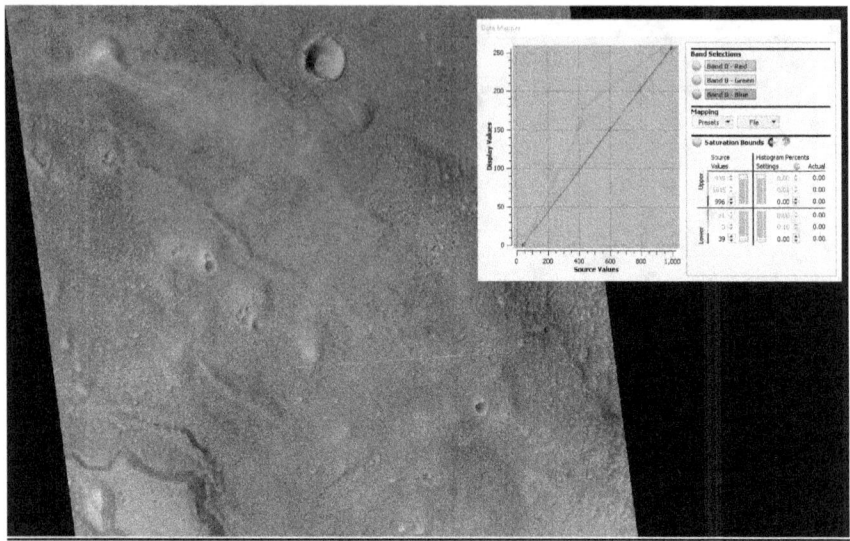

[Figure 52]

Mars Reconnaissance Orbiter picture number: ESP_025505_2210 showing:

Mounds A, B and D circled.

Image acquired on 4th January 2012 by the MRO.

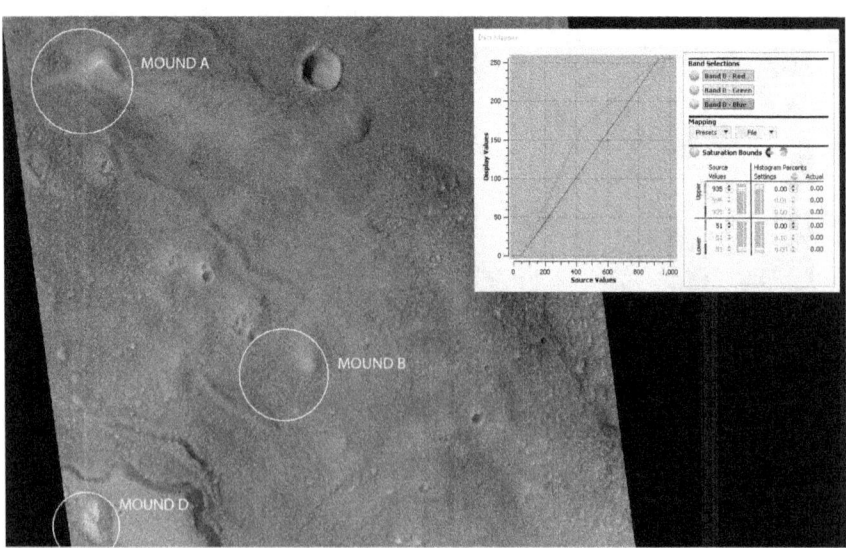

[Figure 53]

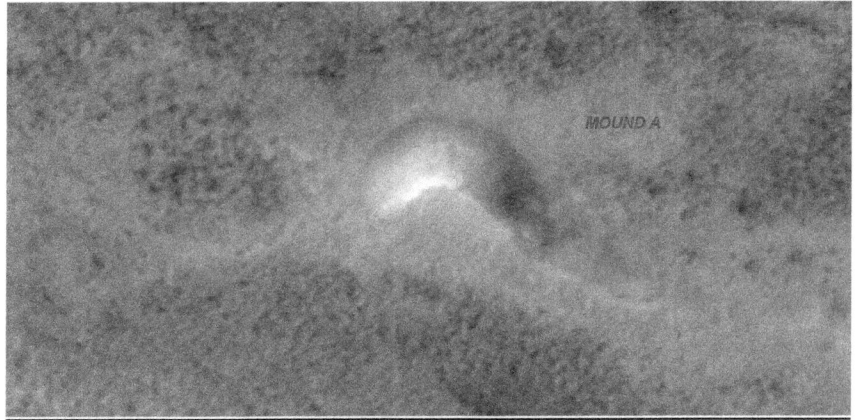
[Figure 54 - Mound A]

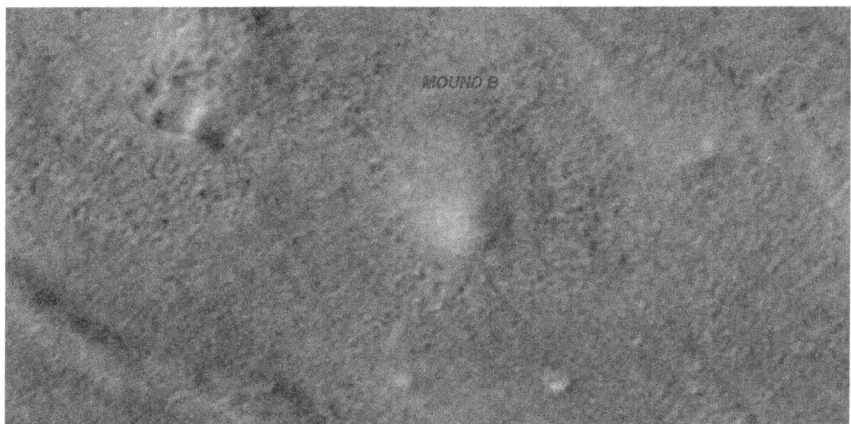
[Figure 55 - Mound B]

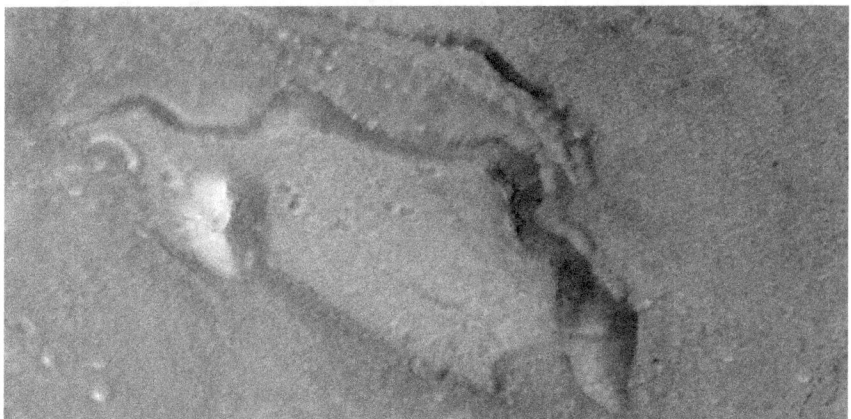
[Figure 56 - Mound D, on a shelf]

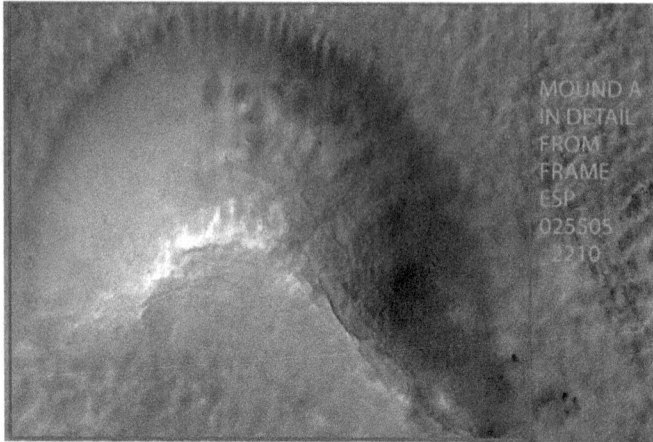

[Figure 57 - MOUND A IN DETAIL]

Mound A, shown above with different lighting from the image shown on the previous page. Frame ESP_025505_2210

Mound B is a very different shape from Mound A. All the mounds display the highly reflective surface tops

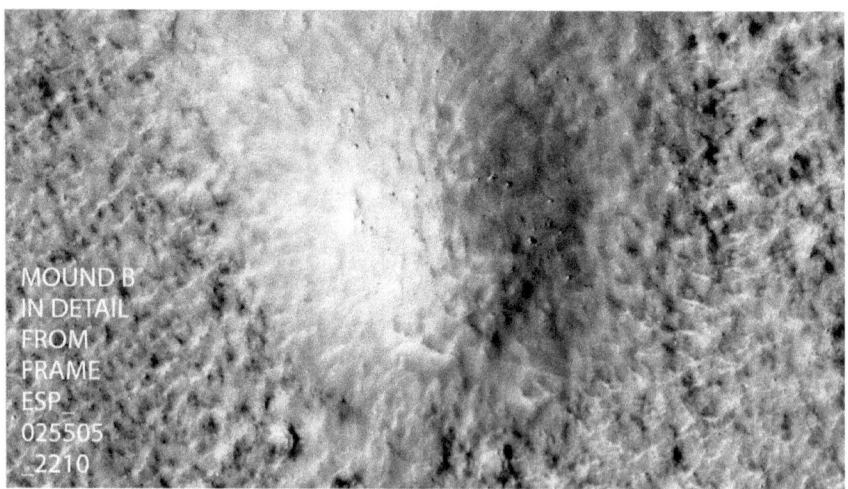

[Figure 58 - MOUND B IN DETAIL]

Let us not forget that there are many mounds around Earth built by humans throughout our history which are of varied shape and size. Some can be construed as gargoyles. Some are burial mounds.

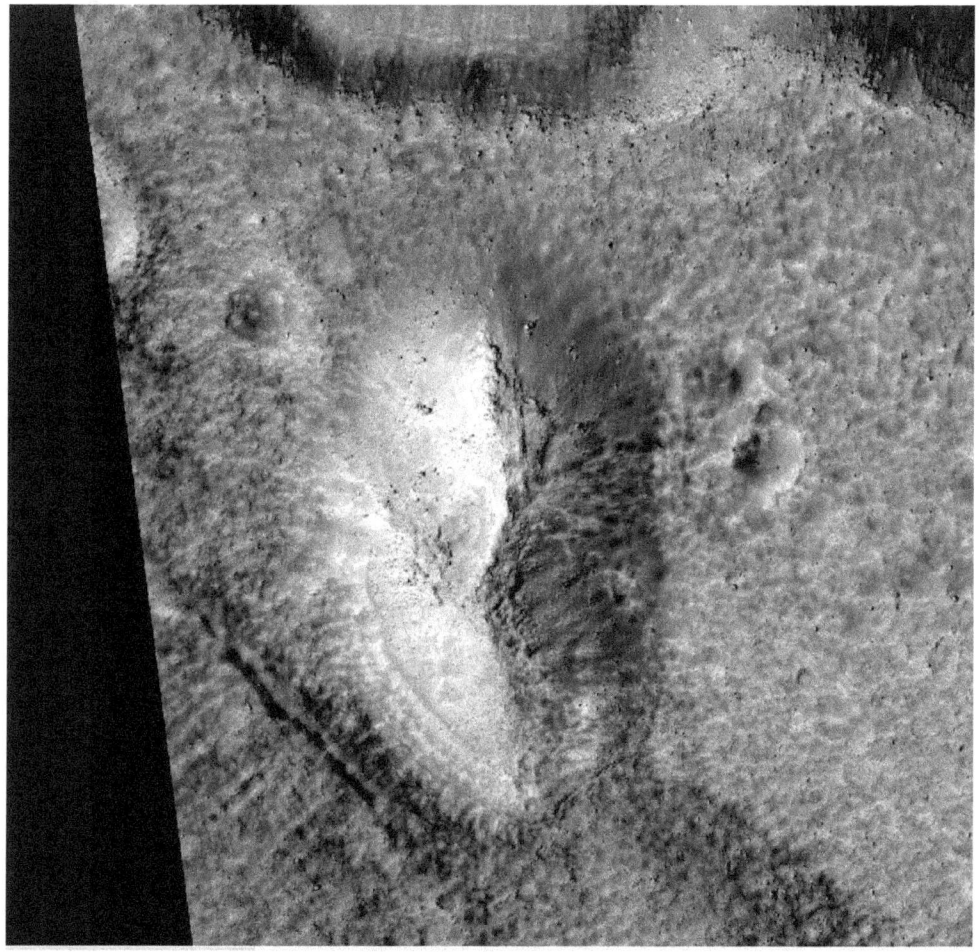

[Figure 59]

Mound D in greater detail - from Frame **ESP_025505_2210.**
The finer details in this image show the shape of this mound clearly.
Picking the centre of the shiny area is difficult at this scale; a fit of the
pentad/hexad pattern was done by Dr. Crater, with all six mounds in the
same picture, to within two pixels. NASA has stated that these images
have a resolution of 5 metres per pixel. Thus, Dr. Crater's selection of the
perceived centres of the mounds is accurate to within 10 metres.

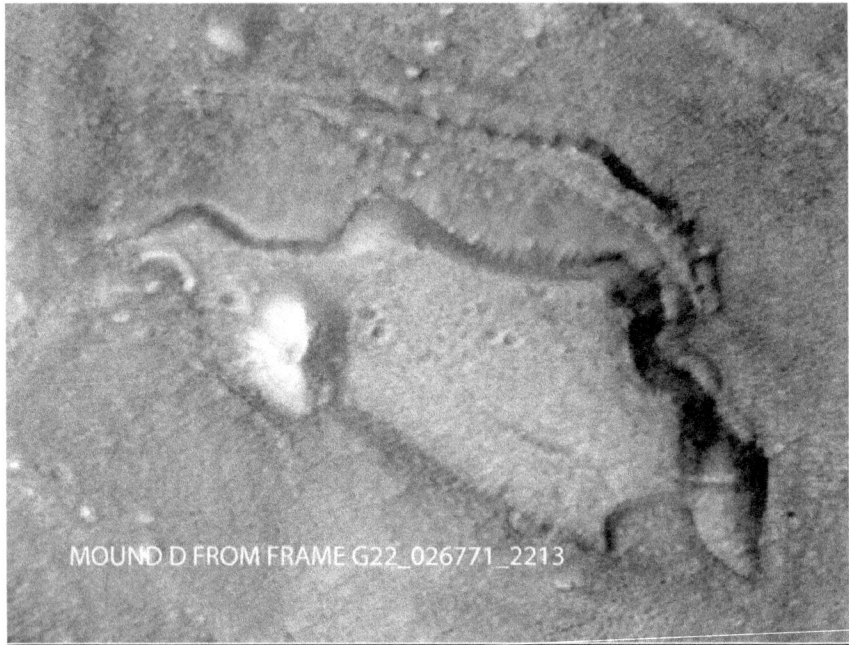

[Figure 60]

Mound D, shown above from Frame **G22_026771_2213**, as a comparison to the previous page's image. The shelf is quite evident here. Mound D has shadows which indicate a different lighting, taken at a different time of day.

The puzzle remains: how did these mounds get positioned into the tight fit displayed by the pentad/hexad pattern?

We do not have an answer to this question - as yet. This is an enduring mystery about Cydonia, an unresolved puzzle on the surface of Mars.

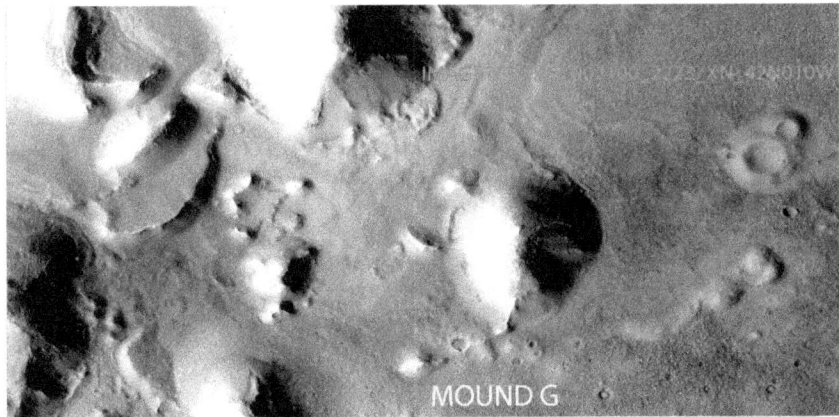

[Figure 61]

Mound G from Image No. P03_002100_2223
Mounds J and K do not form part of this detailed analysis of the hexad even though they appear to fall on the square-root 2 grid lines.

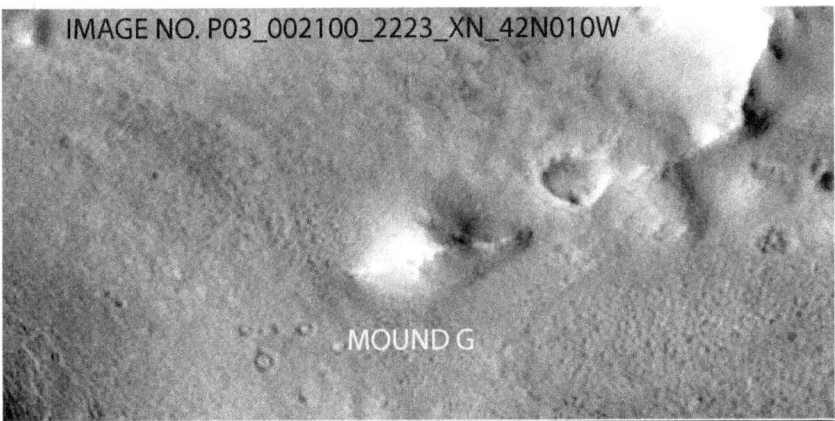

[Figure 62]

Mound G from the same image - enlarged to show an apparent top to its symmetrical shape. The lighting in this image is such that there is a lot of 'washout' due to the shiny surfaces.

This author acknowledges that NASA has released these images without any suppression of data. Their interpretation, analysis and discussion about what they show is open to all researchers.

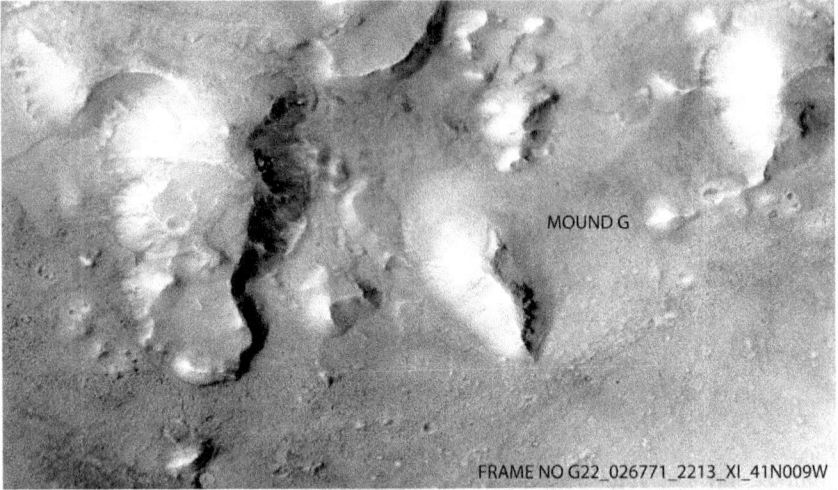

[Figure 63]

Mound G from yet a different frame: G22_026771_2213. There is more 'washout' here than in the previous image. Different sun angles result in different shadows and different reflectivity.

Below is a close-up of Mound G from the same frame.

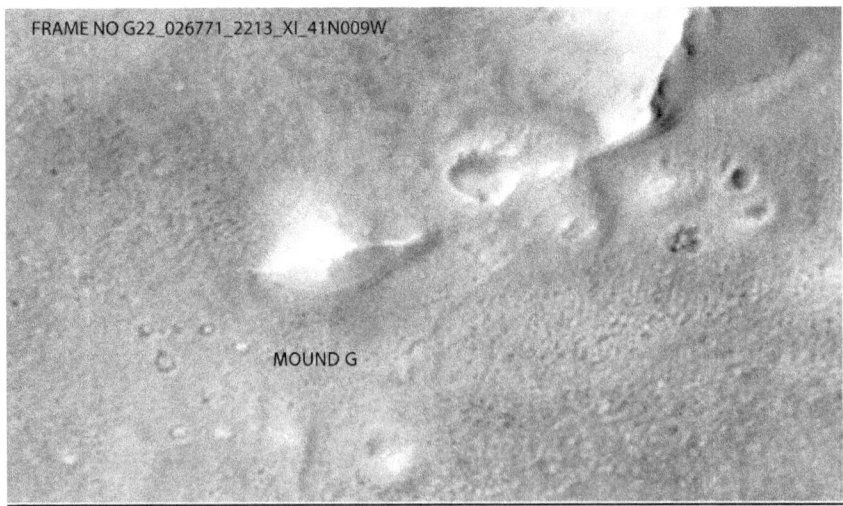

[Figure 64]

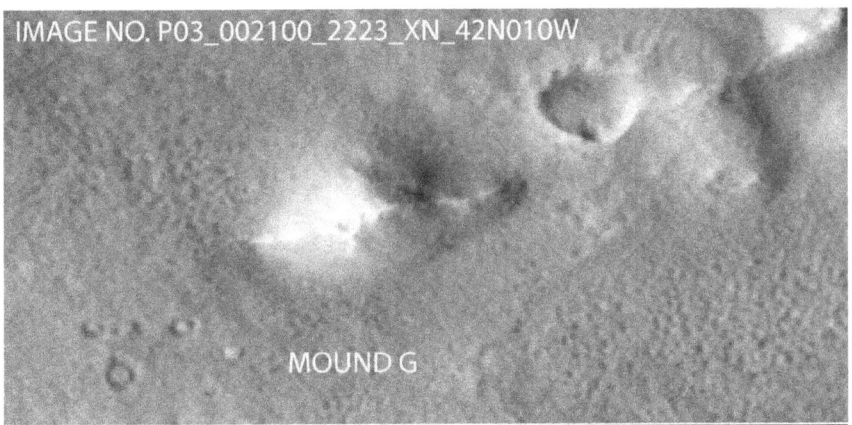

[Figure 65]

Above - Mound G, in different false colour, showing more details.

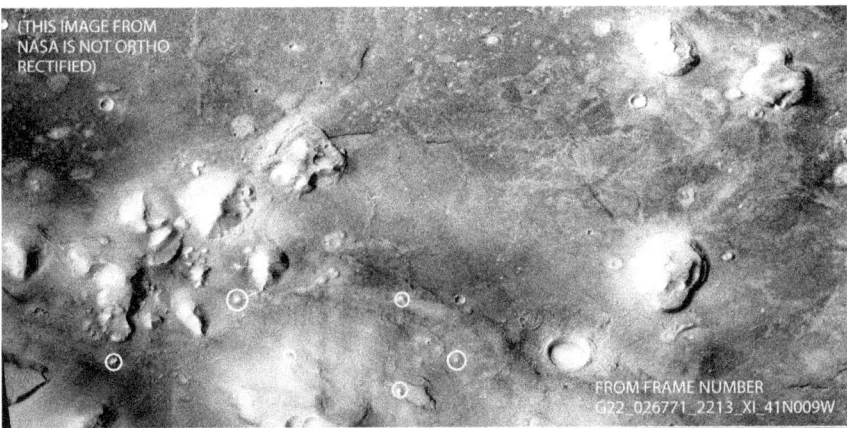

[Figure 66]

The six mounds of the 'hexad' are shown circled above. This is a cropped section of Frame G22_026771_2213, taken by the MRO, a different frame from the previous image. At this scale it is easy to draw lines between the circled mounds. However, the thickness of the lines and the choice of the apparent centre of each mound required a careful positioning of the pentad/hexad to form a fit for the ideal geometry - not a 'forced fit'. Also visible in this image - in the top right quadrant - is the famous "Face On Mars".

[Figure 67]

These images have been released by NASA as non ortho-rectified pictures. Ortho-rectification makes an image appear as though it was taken from directly above the scene, thereby making measurements between features such as the mounds to be more accurate. Most of the measurements given in this book are approximate. Only "ground truth" from the surface of Mars can give extremely accurate distance and size measurements. Future explorers will do surveys in situ, no doubt.

[**NOTE:** There is no Mound C because Dr. Crater dropped use of the letter C in his analysis.]

Picture below shows Mound P in great detail.

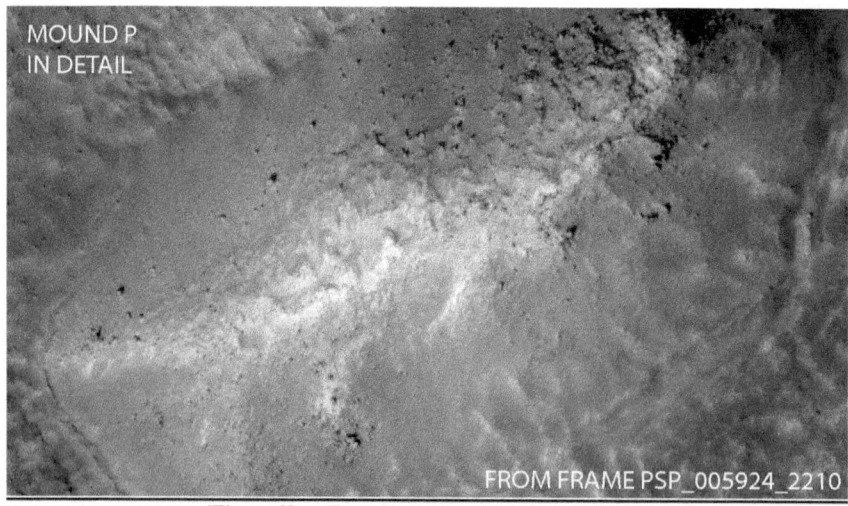

[Figure 68 - From Frame No. PSP_005924_2210]

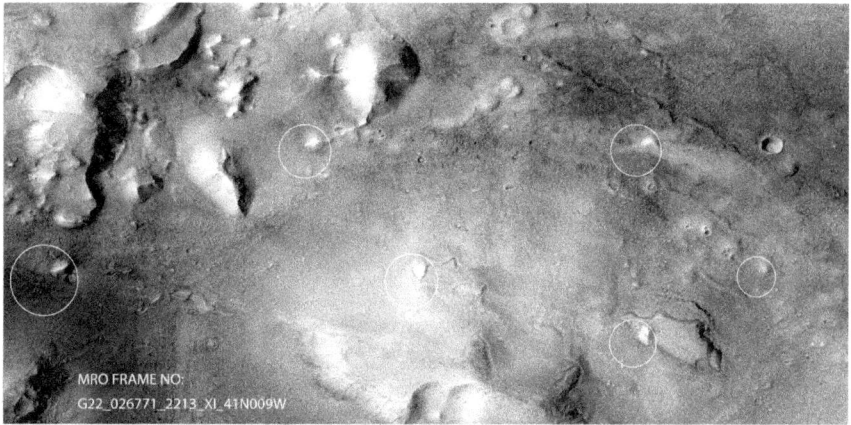

[Figure 69]

The six mounds circled in above enlargement, although not ortho-rectified, still show the pentad/hexad pattern quite plainly. Mound E is shown in detail below. From Frame No. G22_026771_2213.

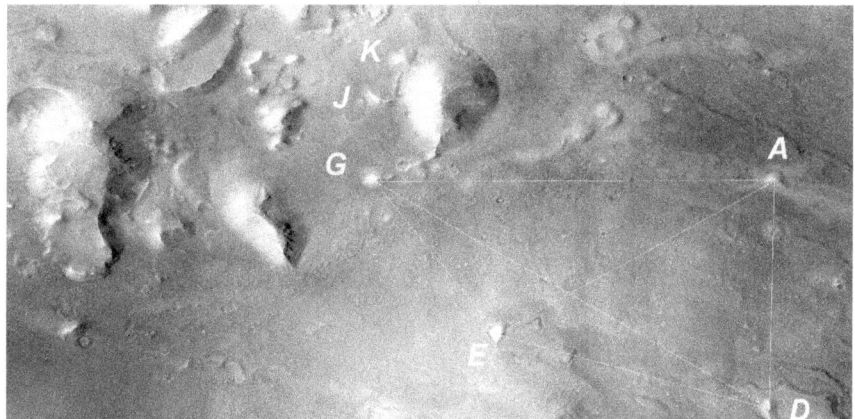

[Figure 70]

Here we see the four mounds GADE (and J, K) from the same image as above, forming the "quatrad". This later confirmation from NASA is welcome; the Viking results of 1976 are confirmed by the later Mars Reconnaissance Orbiter (MRO) camera from the HiRise programme.

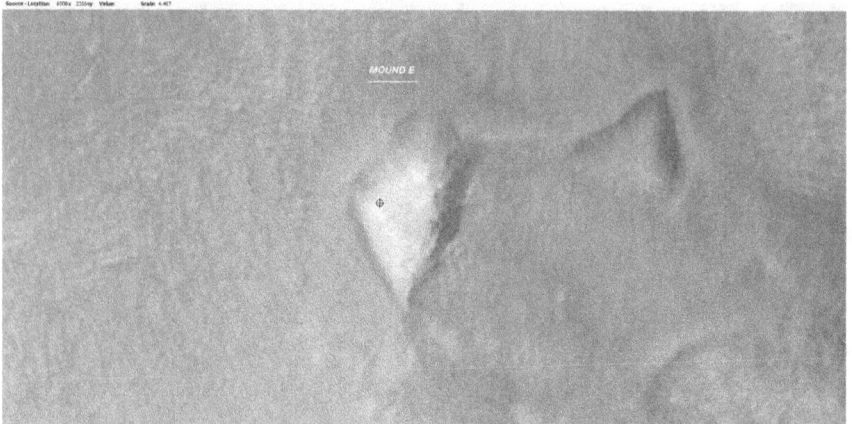

[Figure 71 - Mound E enlarged]

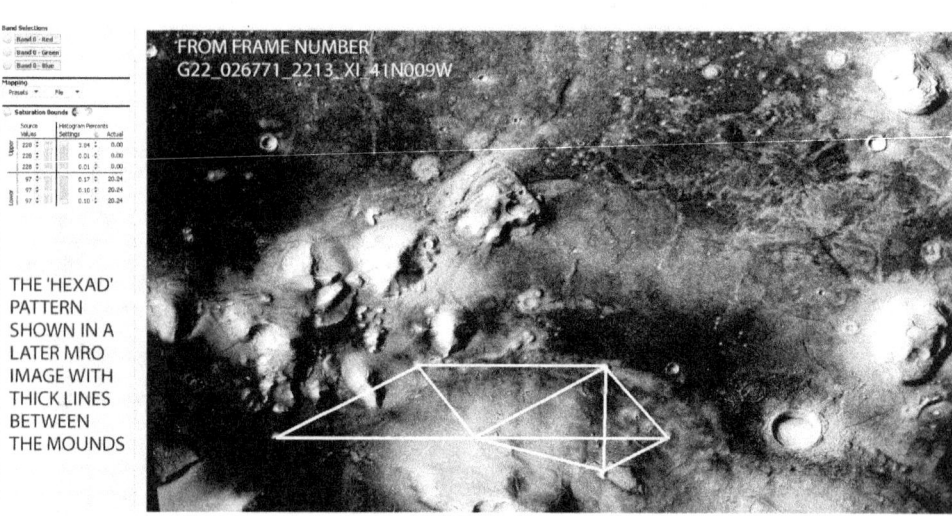

[Figure 72]

Colour and contrast adjustment shows more surface details in this area of Cydonia.

In a book of this size, photographic reproduction naturally is limited to page size. Much greater detail can be evinced from large prints made at the highest quality.

THE SIZES OF THE MOUNDS

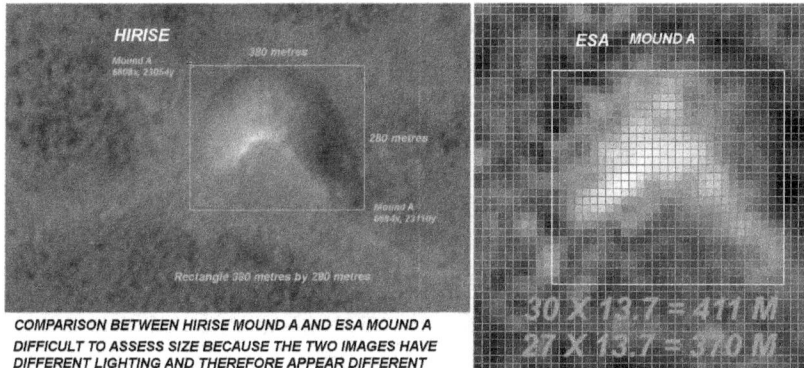

COMPARISON BETWEEN HIRISE MOUND A AND ESA MOUND A
DIFFICULT TO ASSESS SIZE BECAUSE THE TWO IMAGES HAVE
DIFFERENT LIGHTING AND THEREFORE APPEAR DIFFERENT

[Figure 72 a]   [Figure 72 b]

The boxed area around Mound A, from the HiRise MRO image, above left, was calculated to be 380 metres by 280 metres. This assessment was made from the co-ordinates shown on the computer screen in the x,y directions. NASA stated that the resolution of the HiRise image is 5 metres per pixel - the rectangular box drawn around Mound A is shown to be 76 x 56 pixels in size:

$$76 \times 5 = 380 \text{ metres}$$
$$56 \times 5 = 280 \text{ metres}$$

The same mound, above right, in comparison, shown under different lighting conditions by the Mars Express camera from the European Space Agency (ESA) was estimated to be 411 metres by 370 metres, depending on how many square pixels in the image was counted inside the box. ESA declared the resolution of their images of Cydonia to be 13.7 metres per pixel and to have been processed before release to the public.

The resolution of the 1976 Viking frame 35A72 was 47 metres per pixel.

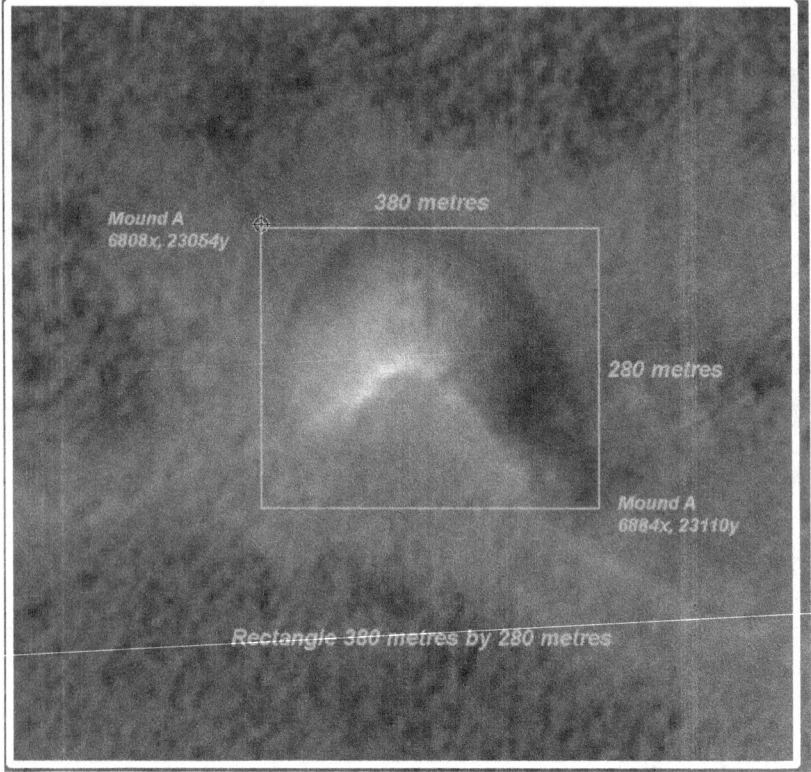

[Figure 72 c]

The difficulty of assessing the precise size of the mounds is evident from the above enlargement. However the shiny top part of each mound is visible in all photographs of the area. To repeat, the mounds are of different shape and their sizes fall into a bracket of measurements.

The accuracy of measurement is always specified in any case; example to the nearest inch or nearest foot, or nearest mile. At a microscopic level naturally the brackets of measurements are much smaller in scale, to fractions of inches or millimeters.

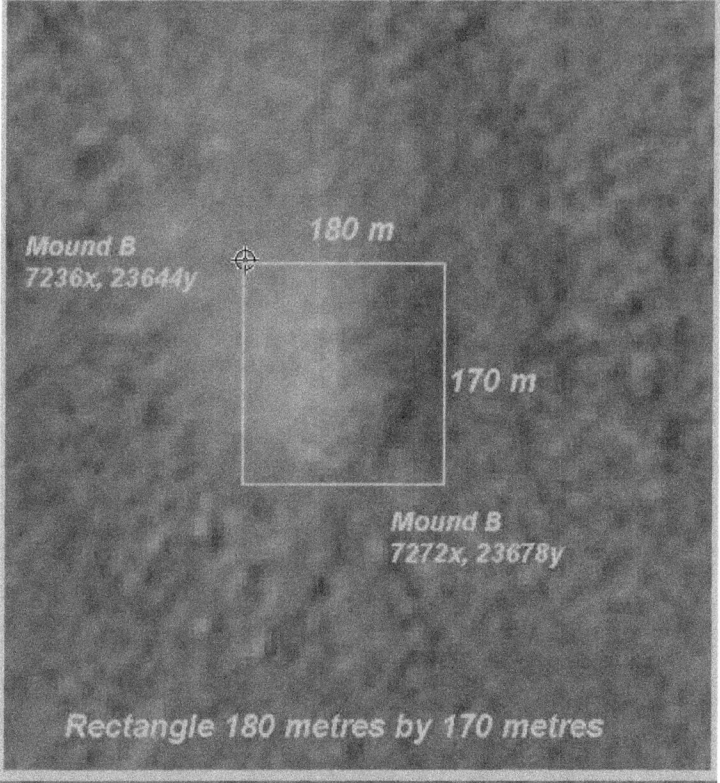

[Figure 72 d]

Mound B was boxed into a rectangle 180 metres by 170 metres and was therefore estimated to be the smallest of the six mounds of the hexad. Some have remarked on the strange appearance of this mound.

As with all observation, the closer one gets to an object, its overall shape and size become lost in finer detail. There is an optimum distance for all viewing, depending on the size of an object.

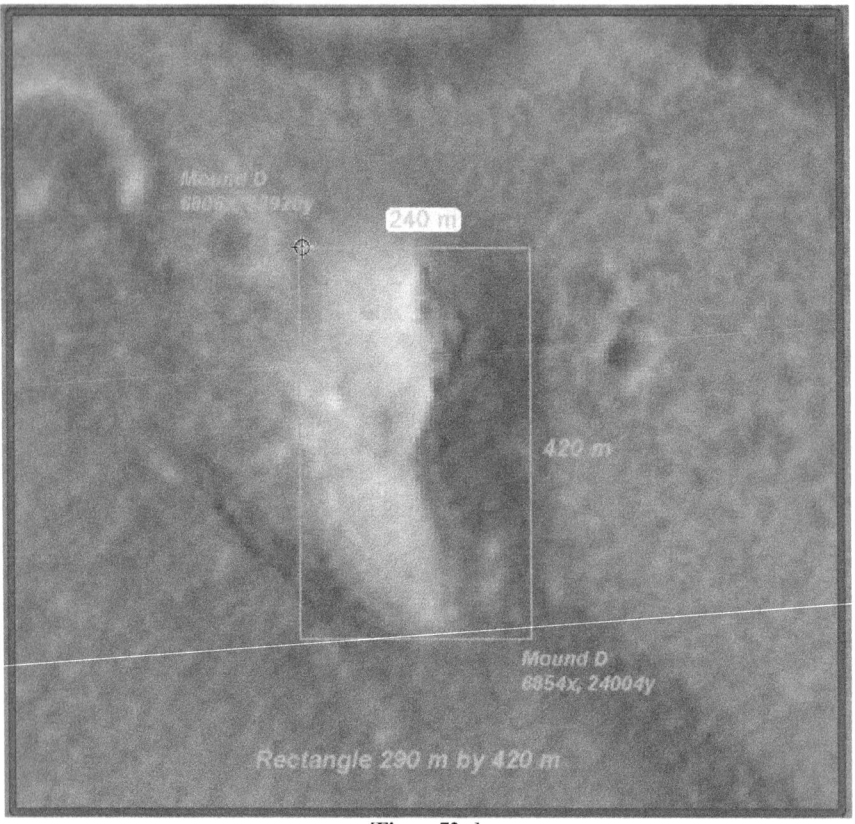

[Figure 72 e]

Mound D was judged to be about 240 metres by 420 metres with a high albedo summit.

If all of these mounds had been square in shape, with a clear top, such as the pyramids of Egypt or south America, it wiuld have been easier to pinpoint their peaks.

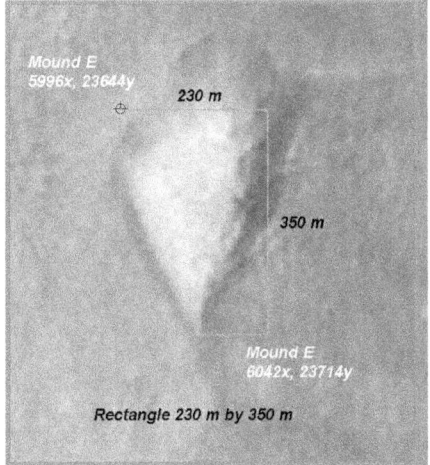

[Figure 72 f]

Mound E estimated to be 230 metres by 350 metres.

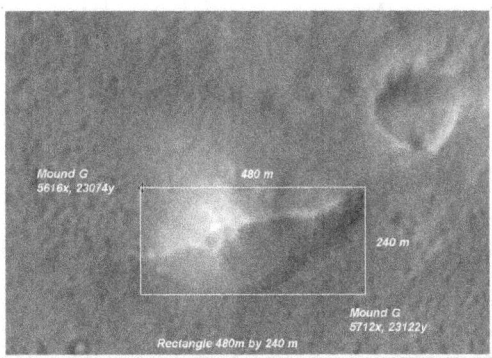

[Figure 72 g]

Mound G estimated to be 480 metres by 240 metres.

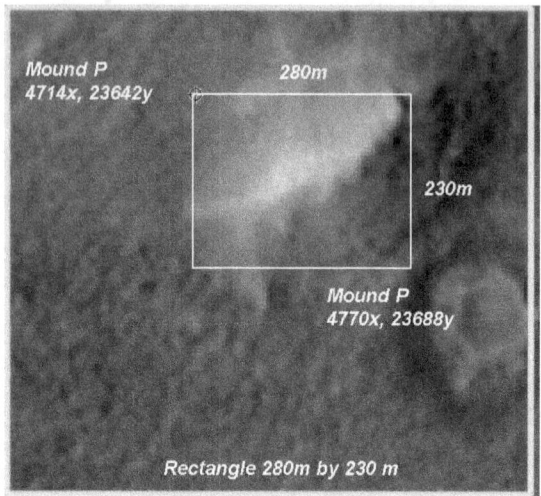

[Figure 72 h]

Approximate sizes of Mounds in Cydonia

MOUND	LENGTH OF BOX	HEIGHT OF BOX	AREA IN SQ metres
A	380	280	106,400
B	180	170	30,600
D	240	420	100,800
E	230	350	80,500
G	480	240	115,200
P	280	230	64,400

[Table 2]

**Pyramid at Giza in Egypt has a square base of 230.34 m
Area = 53,056.5 square metres**

**Thus we see that the Giza pyramid area lies between that of area of Mound B and Mound P.
The boxes in all Figures above were selected by this author.**

[Figure 73]

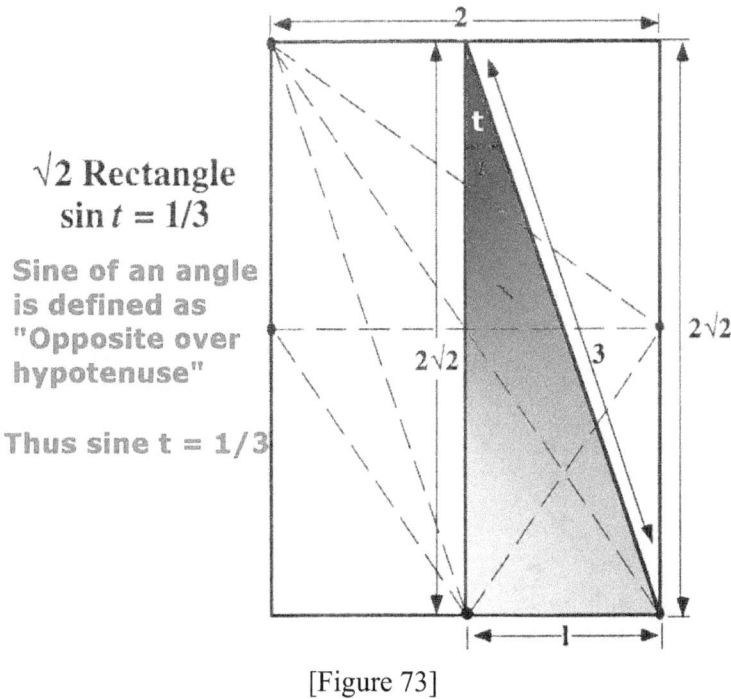

[Figure 73]

(Reminder: angle t = 19.47.....degrees)

A simple reminder of a basic rule in trigonometry - the definition of the "sine" of an angle - "opposite over hypotenuse".
From the same triangle, the "cosine" of angle t would be "adjacent over hypotenuse", 2root2 over 3; i.e. 2root2 / 3.
Likewise, the "tangent" of angle t would be "opposite over adjacent", giving us a value of 1 over 2root2 (1/2root2)

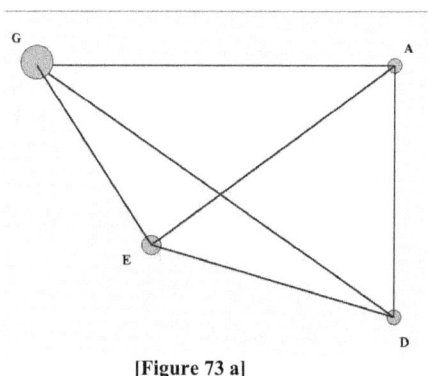

[Figure 73 a]

THE MOUND CONFIGURATION IN CYDONIA - AN UNRESOLVED MYSTERY ON MARS

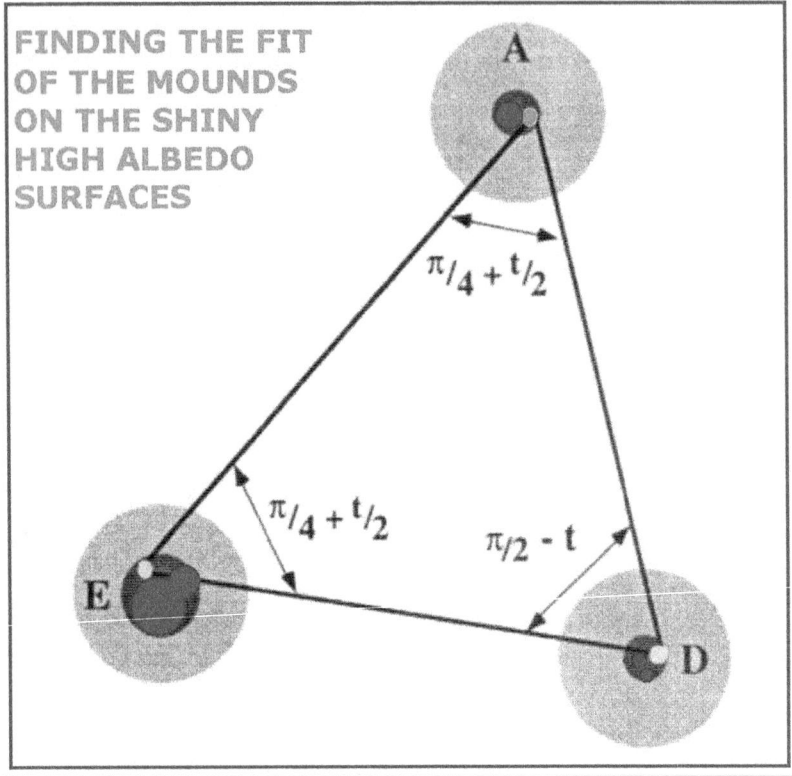

[Figure 74]

The figure above shows the vertices of triangle EAD in a fit inside a circular spot on each mound. Dr. Crater worked to one pixel on the original Viking image with the shiny spots represented by the smaller, dark circles.

Obviously such a fit would be very easy if the lines between the mounds were extremely thick. Dr. Crater however required the entire fit over the 'hexad' layout to be done with the thinnest possible lines, a reflection of his desire to work within a pixel or two. The Viking frame 35A72 had a resolution of 47 metres per pixel. Later images had better resolution.

On average, the area covered by each mound is about the same or larger than the great Khufu pyramid at Giza. The three angles shown above: **l, r, r** are represented by pi and t. = Π) + **t/2**

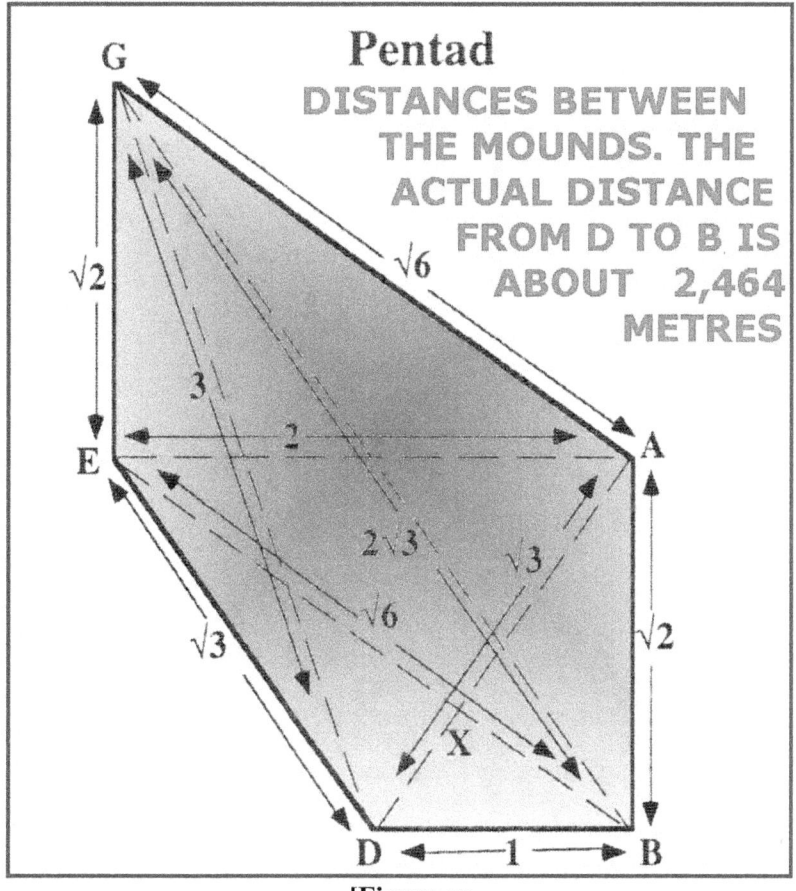

[Figure 75]

If:

DB = 1 unit (actual 2464 metres; 1.53 miles) **then**
BA = root 2 (3484 metres or 2.16 miles)
DE = root 3 (4268 metres or 2.65 miles)
EA = 2 (4928 metres or 3.06 miles)
GE = root 2 (3484 metres or 2.16 miles)
GA = root 6 (6035 metres or 3.75 miles)
GB = 2 root 3 (8536 metres or 5.30 miles)
GD = 3 (7392 metres or 4.59 miles)
DA = root 3 (4268 metres or 2.65 miles)

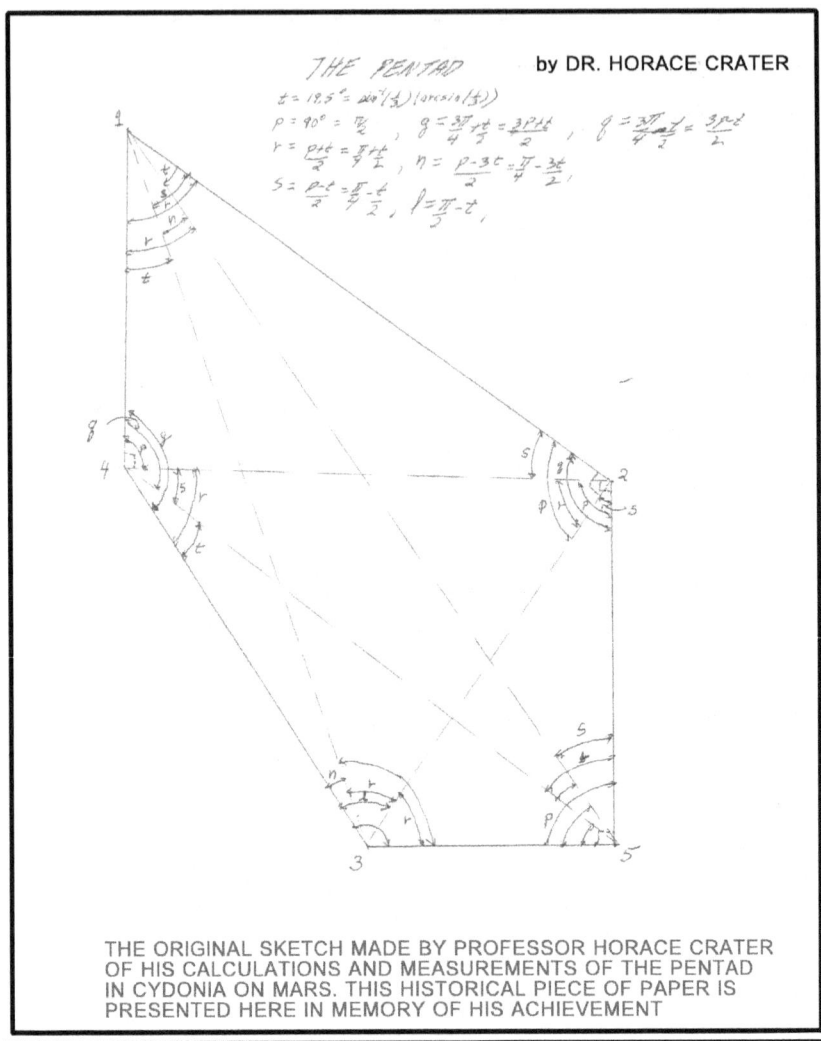

[Figure 75 a]

The original sketch made by Dr. Horace Crater had the word "breathless" written on it. This was his indication of how he had to control his excitement as the ground plan on Mars unfolded. The above piece of paper is in his own handwriting.

AREAS OF THE TRIANGLES IN THE PENTAD

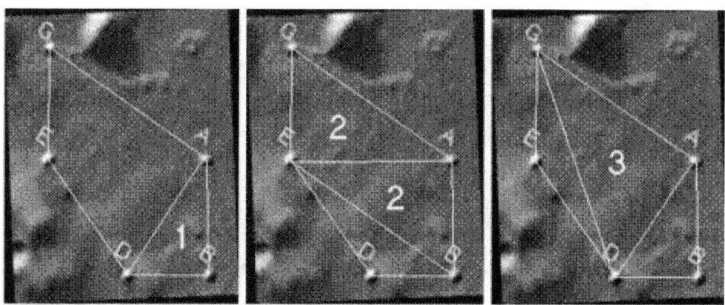

[Figure 76]

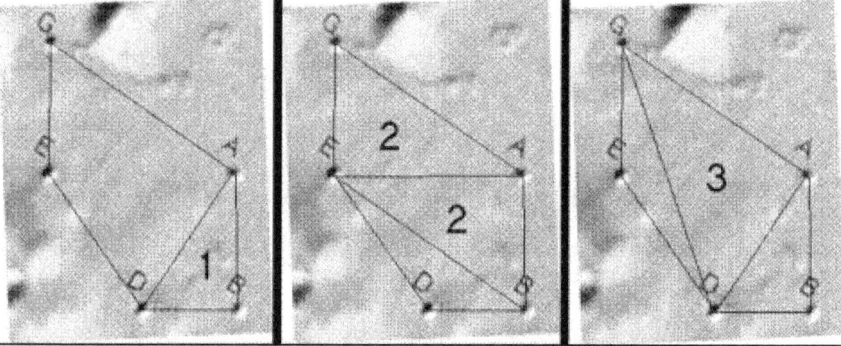

IF AREA OF ABD = 1 THEN AREA OF EAB & GEA = 2 AND AREA OF GAD =3
[Figure 77]

If the area of triangle ABD was said to be 1 unit, then the other triangles shown here would be 2 or 3 units in size, whatever the unit is declared to be. The actual area of triangle ABD is approximately:

Half of: [2464 m x 3484 m = 8,584,576 square metres]
   (1/2 of 8.585 = 4,792 square kilometres approximately)

Thus: Area of ABD = 4,792.5 square km

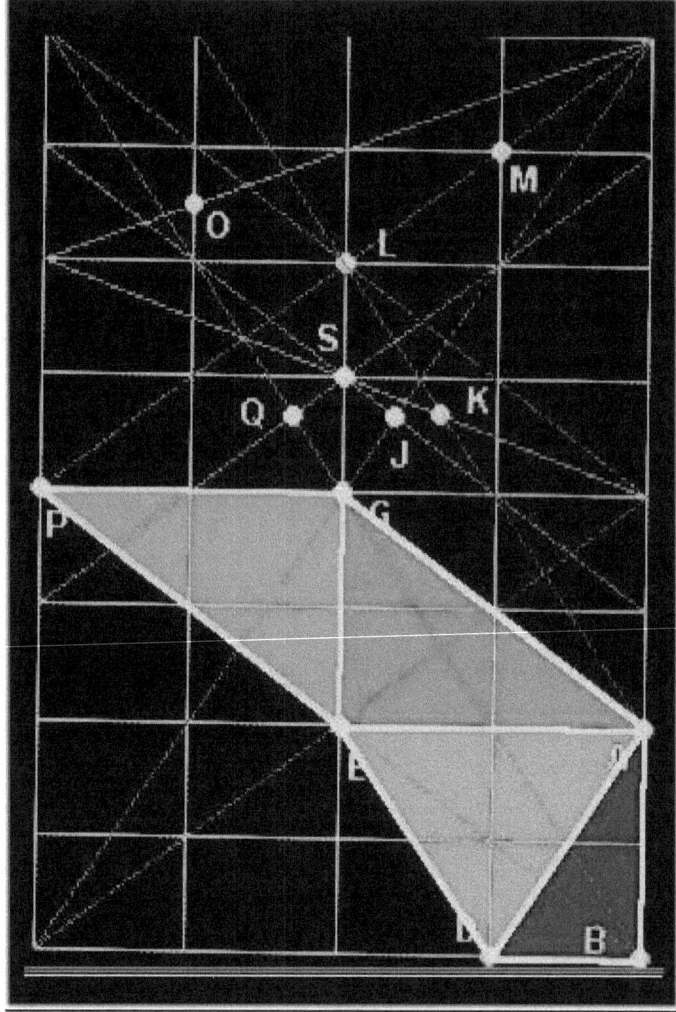

[Figure 78]

So, in the above figure, if the blue triangle was stated to be of area 1 unit, then the red and green triangles would be of area 2 units each.
Triangle GAD is not outlined in the above drawing of the hexad but its area would be 3 units, as per the previous page.

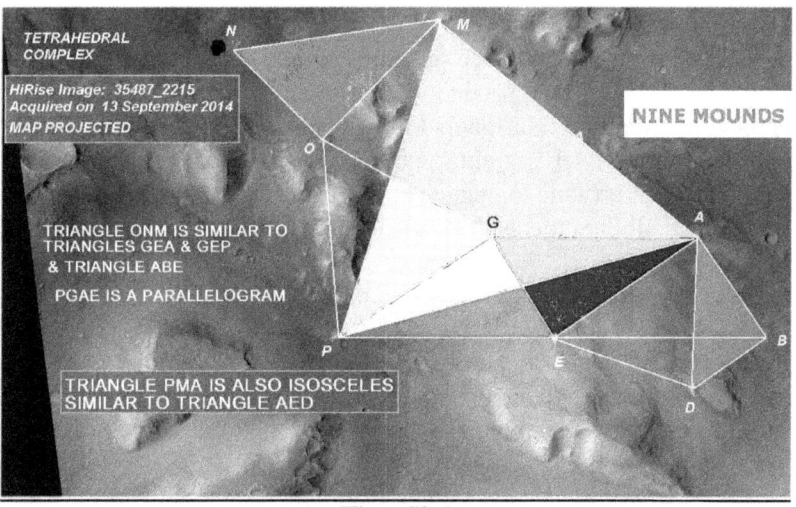

[Figure 78 a]

If anything is evident from the above image, showing 9 mounds, **is the apparent organisation in their layout.**

Triangle PMA is isosceles, similar in angles to triangle AED.
Triangle OPG appears to be equilateral.
Triangle ONM is displaced from triangle GEA by about 11 degrees.

Note again - Dr. Horace Crater limited his analysis to those mounds similar in size and area and did not venture to study the very large mounds. Other researchers are urged to look at those mounds too.

One researcher who has looked at the larger mounds has posited a pentagon surrounding the four smaller mounds. Architect Robert Fiertek has presented his plan here:

https://www.bibliotecapleyades.net/marte/esp_marte_17d.htm

Fiertek wrote:

"Design theories exist to mirror the will of the society, reflecting that society's intent. Compare our modern architecture's emphasis on clean planes to simplify with the medieval cathedral builders' efforts at vaulting the spaces to impress on visitors a visceral sense of the presence of the Divine. "The basic idea of the Cydonia complex seems to be based in the celebration of musical and mathematical patterns.

"Let us assume, as the evidence seems so strongly to suggest, that the Cydonian City is indeed a product of intelligent design. A detailed examination reveals the designers to have been quite facile at expressing complexity without delving into randomness. They repeatedly chose to let design factors intertwine to engage and influence the design (as opposed to having had all the objects the same shape or size or orientation).

"This web of geometric interrelationships among structures betrays deliberate design in many basic ways. By further interpreting this evidence it may eventually be possible to see into the mindset of the creators of these structures, through the degree and quality of complexity of design, and to gain some insight into their sensibilities.

"The complexity in the designs and the ability of the designers to transcend simplistic clichés is far greater than that currently being generated by design theory on Earth. "   -   Robert Fiertek

---

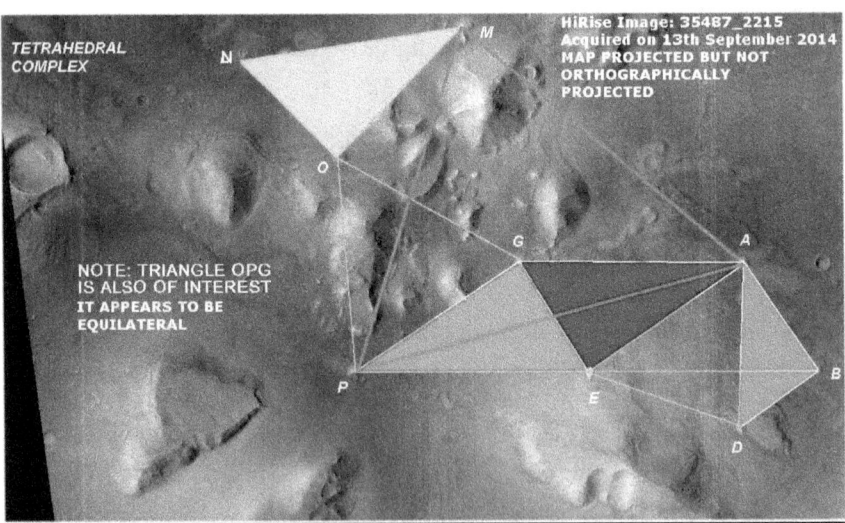

[Figure 78 b]

Reminder: PGAE is a parallelogram. It cannot be defined as a rhombus because its four sides are not equal in length.

What exactly has been going on in Cydonia? Is this pattern a result of nature's geological forces at work? Or can we suggest intelligent layout?

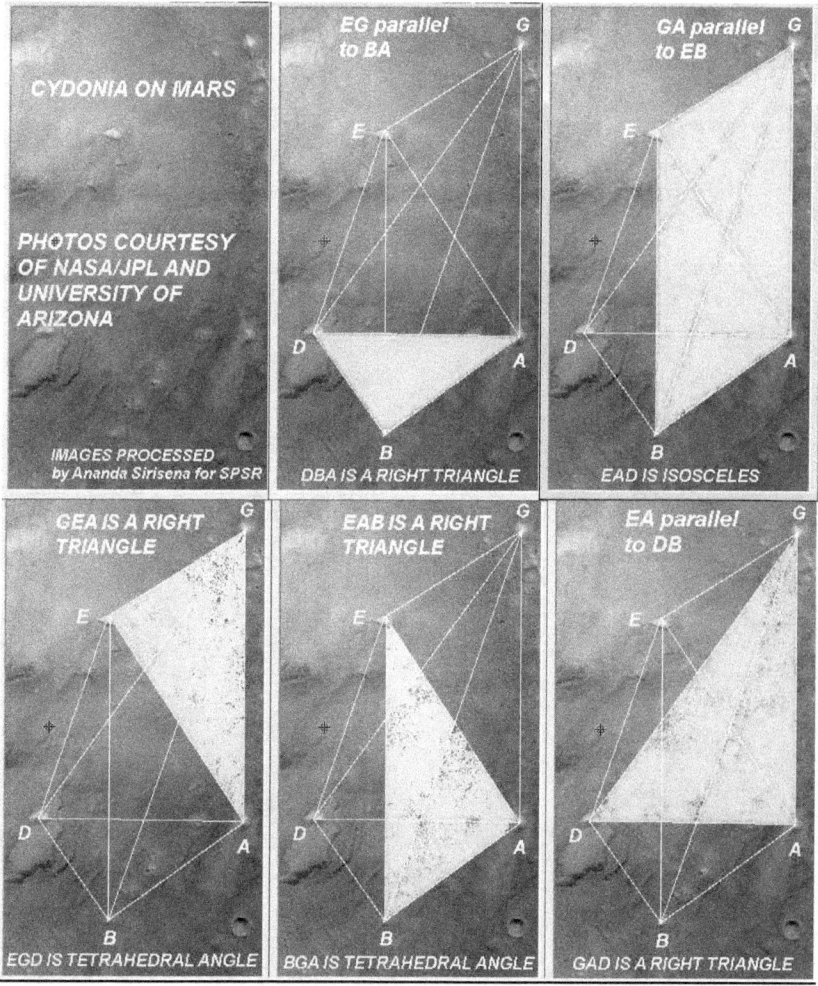

[Figure 78 c]

The pentad - rotated so that line GA is in a north-south direction. The number of right-angled triangles is evident even in the "pentad" above. The "hexad" shown on the previous page also shows sets of parallelograms, in addition to other triangles indicative of a planning venture.

Dr. Crater wrote: "A **co-ordinated fit** requires that the same vertex within any given mound is used for all the triangles having one vertex sharing that mound, not shifted about arbitrarily within each mound to accommodate each triangle separately."

Chapter 5

CYDONIAN MOUND GEOMETRY:- A Closer Look

[Figure 78 d]

by Professor
STANLEY V. McDANIEL

Copyright © 1996 by Stanley V. McDaniel, emeritus professor
Reproduced here with the permission of professor Stanley V. McDaniel

Abstract

In Dr. Horace W. Crater's analysis of the mound configuration located in the Cydonia area of Mars it was pointed out that a group of five small, relatively distinct and uniform features referred to as mounds labelled GABDE correspond to the positions of five nodes (corners and side midpoints) of a classic geometric figure known as a "square root 2 (sqrt2) rectangle.

Note The present paper presents a detailed account of the geometric relationships between the mound pentad, the geometry of the tetrahedron, and the geometry of the sqrt2 rectangle. **A consideration of possible cultural or symbolic meaning for this geometric formation is also presented.**

1. Introduction

Is the investigation of geometric relationships between objects on another planet in order to evaluate their possible artificiality a legitimate endeavour?

One scientist has stated that since geologists would never consider making such "planimetric measurements," there is no point in anyone else doing so. But when questions of possible cultural or symbolic meaning arise, geology is not the relevant science. Rather it is those disciplines which study systems of cultural symbolism and other manifestations of culture -- archaeologists, anthropologists, and experts in the communication of ideas by means of symbols.

Geometry provides a universal "language" for the expression of both aesthetic and symbolic ideas. There is no clear reason why that potential universality should not apply to extraterrestrial intelligence as well as Earthly cultures. In searching for a possible meaning of the mound configuration at Cydonia, it is relevant and appropriate to explore the geometric and possible symbolic characteristics of the mound configuration.

[NB; The diagrams in this chapter are given the notation Appendix Diagram 1 to Appendix Diagram 12. The author extends his deep gratitude to professor Stanley McDaniel for granting permission to reproduce this chapter.]

Questions that bear on the issue of artificiality are the following:
 Does the mound geometry conform to some understandable grid?
 Is the mound geometry commonplace or unique?
 Is the mound geometry unproductive or rich?
 Is there any recognizable cultural or symbolic quality in the
patterns that might provide a clue to possible meaning or utility?

In the discussion below, we will try to answer these questions.

2. The Tetrahedron Cross-section

The two types of triangles that were found to generate a radical statistical anomaly among the 12 mounds at Cydonia are those containing angles related to the cross-section of a tetrahedron. We begin with a diagram showing how the cross-section, which is an isosceles triangle, appears within the tetrahedron.

Tetrahedral Cross-Section

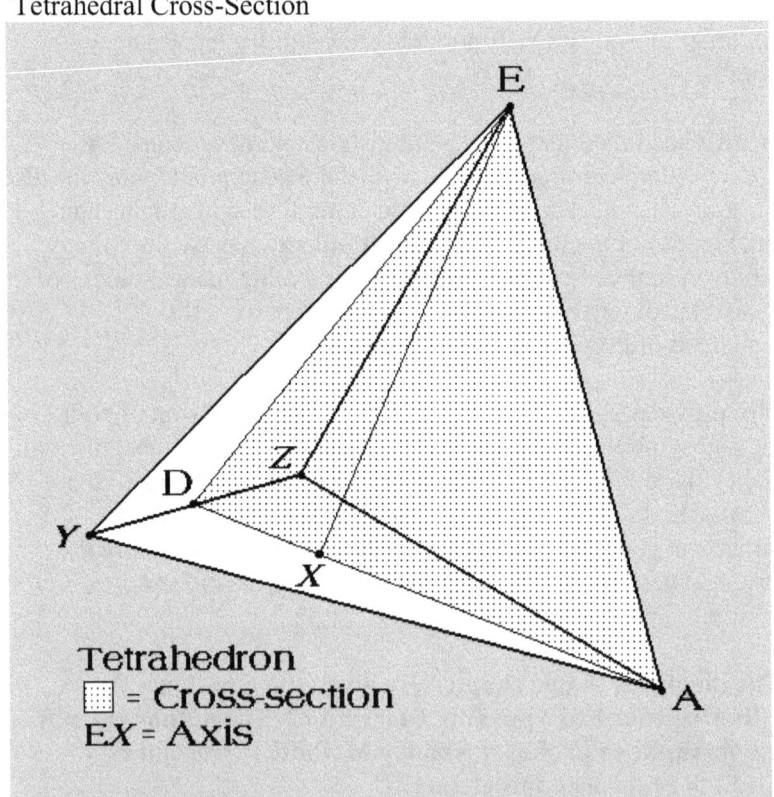

[Appendix Diagram 1]

Mounds EAD mark out a triangle matching the cross-section (shaded area) of tetrahedron EAYZ.

In this diagram, line EA is one edge of the tetrahedron with point E at the apex. Point D lies at the midpoint of the opposite edge, and line AD bisects the base of the tetrahedron. Line EX represents the axis of the tetrahedron. Point X is equidistant from A, Y, and Z on the base of the tetrahedron. Line EX is also an altitude of the cross-section as well as the axis of the tetrahedron. (Altitude = line from any vertex drawn perpendicular to the opposite side.)

If we draw all the altitudes of the cross-section we find that it is divided into triangles having certain characteristic geometric properties. Let us look first at the two triangles formed within the cross-section by altitude EX.

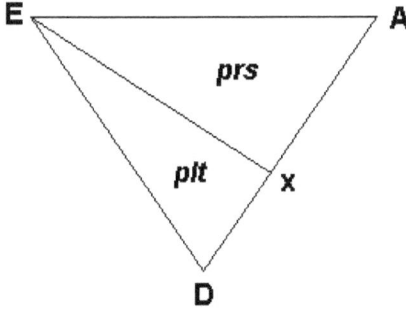

Appendix Diagram 2

Altitude EX divides the cross section into prs and plt triangles.

This basic division, using the axis of the implied tetrahedron, divides the cross-section into triangles we have termed "prs" and "plt" triangles with the following internal angles to be designated p, t, l, r, and s (see diagrams 5a,b and 6a,b below for the exact location of each of these angles within the triangles):

Abbreviation	Degrees	Radians
p	90	pi/2
t	19.5	0.34
l	70.5	(p - t)
r	54.75	(p + t) / 2
s	35.25	(p - t) / 2

All the angles are simple functions of pi and another angle t = 19.5 degrees (0.34 radians), which turns out to be a tetrahedral constant, as will be explained below.

Footnote These triangles have specific roles to play in this particular geometric form. Let's take a look at what happens when you divide the cross-section into sub-triangles by drawing all three altitudes.

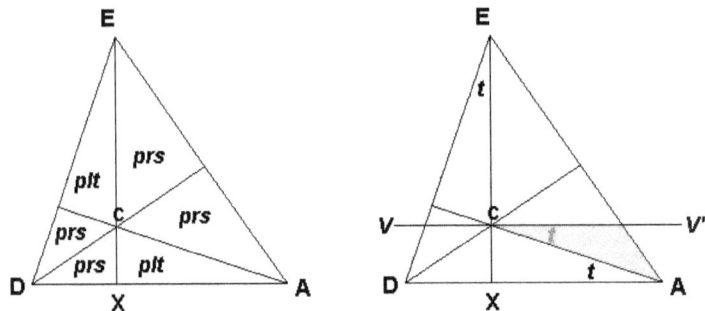

Appendix Diagrams 3a (left) and 3b (right)

Division of the cross-section by all its altitudes, and the geometric result of a line drawn through the center parallel to the base.

When we divide the cross-section by all three altitudes, multiple instances of prs and plt triangles spring into being, the precise geometric center of the tetrahedron (at C) is identified, and one specific angle -- the tetrahedral constant "t" -- takes on an important role. Because the angle formed by the altitude at A equals t, line V-V' drawn through the center point parallel to AD forms another instance of the angle t (shown in red). Below we will explain the geometric consequences of this fact.

3. The Tetrahedral Latitude

The presence of angle t at V'CA is of considerable geometric significance. It determines the value known as the "tetrahedral latitude" (19.5 degrees). This is the latitude where the base of a tetrahedron, when embedded within a sphere, meets the surface of the sphere (taking the apex of the tetrahedron as a pole of the sphere). Diagram 4 below graphically demonstrates how this comes about.

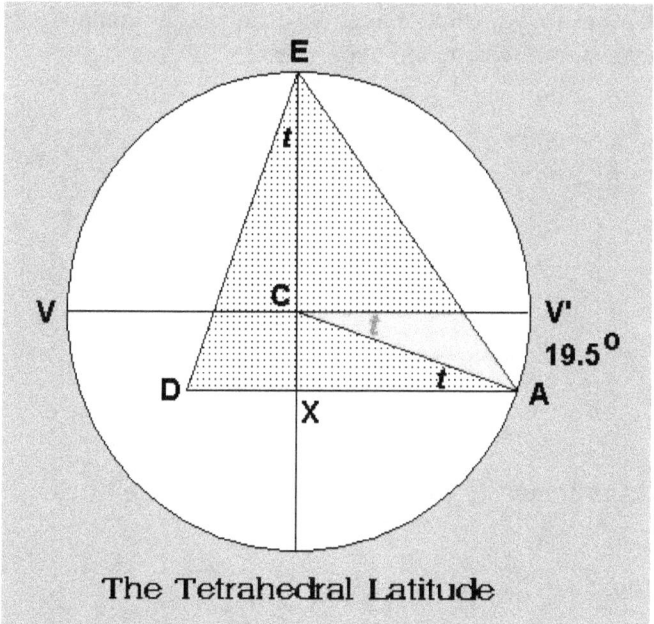

Appendix Diagram 4

Cross-section of a tetrahedron embedded within a sphere, showing the tetrahedral latitude. (Compare Diagram 3b)

Because this is a cross-section of both the sphere and the embedded tetrahedron, it may be a bit tricky to visualize the full three-dimensional figure. Point D is not a vertex of the tetrahedron, but a point within the sphere on an edge of the tetrahedron that runs at right angles to the screen you are viewing this on. Compare Diagram 4 with Diagram 1 to see where point D is located. In Diagram 4, if you were to draw a line perpendicular to the computer screen directly through D, where this line meets the surface of the sphere (in front of and behind the screen) would mark the other two vertices of the tetrahedron (at Y and Z in Diagram 1).

4. How the Sqrt2 Rectangle is related to the Tetrahedron Cross-Section

When an altitude is drawn in the cross-section from the large angle at D, this divides the cross-section (which is an isosceles triangle) into two equal right triangles. Each of these right triangles turns out to be our familiar prs right triangle. Suddenly a geometric relationship between the cross-section and the sqrt2 rectangle springs into view: The sqrt2 rectangle is simply the cross-section of a tetrahedron with its component prs triangles moved to share a common hypotenuse.

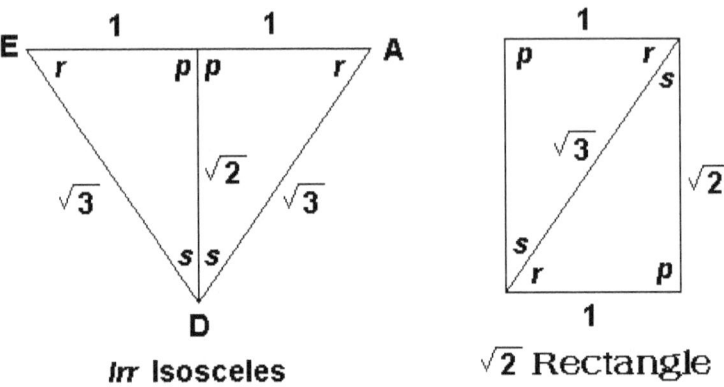

Appendix Diagram 5a (left) and 5b (right)

In 5a the lrr isosceles triangle is divided by its altitude into two prs right triangles. In 5b these same two triangles share a common hypotenuse to create a sqrt2 rectangle.

Diagram 5a above illustrates another important aspect of the rich geometry of the tetrahedron cross-section. When the half-base is taken as 1 unit, the altitude of the triangle is sqrt2 (= 1.414) and the sides are sqrt3 = (1.732). This geometry is what establishes the relationship between the tetrahedron cross-section and the sqrt2 rectangle. In the sqrt2 rectangle, the short side = 1, the long side = sqrt2, and the diagonal = sqrt3. Thus all that is needed to produce the sqrt2 rectangle is to place the two prs right triangles found in the cross-section so that they share a common hypotenuse, as shown in Diagram 5b.

5. Rich Geometric Properties of the Sqrt2 Rectangle

Investigation of this figure shows that it is geometrically unique. Besides its relation to the tetrahedron cross-section shown above, It is the only rectangular figure which, when divided in half by a line perpendicular to its long side, yields smaller equal rectangles having the same proportions as the larger figure. The only other plane geometrical figure with this property is the 45 degree right triangle.

Because of this unique property, when the rectangle is divided in half first horizontally then vertically, four smaller rectangles with the same proportions are created. Since this process could go on indefinitely with the same proportions being maintained in smaller and smaller segments, the sqrt2 rectangle appears to have what one might call a holographic quality, which easily lends itself to symbolic interpretation. For example, the process of cell division, and in general the repetition of form in various natural processes, might be symbolically represented by this geometrically self-replicating figure.

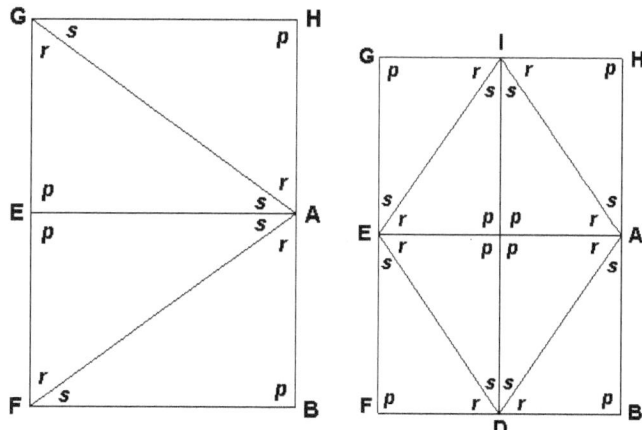

Appendix Diagram 6a (left) and 6b (right)

The sqrt2 rectangle reproduces its proportions ad infinitum when bisected perpendicular to its long axis.

6. Representation of the Tetrahedron Cross-Section in the Sqrt2 Rectangle

When lines are drawn between all eight nodes of the rectangle, what we see are multiple repetitions of the tetrahedral cross-section. In Diagram 7a below, There are two very large cross-sections (GAF and HEB), four more 1/2 the size of the larger ones (for example EAD), and four more 1/2 that size again (with their bases at GE, EF, AB, HA). There are also multiple repetitions of prs and plt triangles. To make these stand out more clearly, Diagram 7b differentiates the various triangles using color coding. (There are many more plt triangles than just the ones shown in red. If you extend the lines of the red triangles you will see further versions of the plt triangle appearing at every other crossing line.)

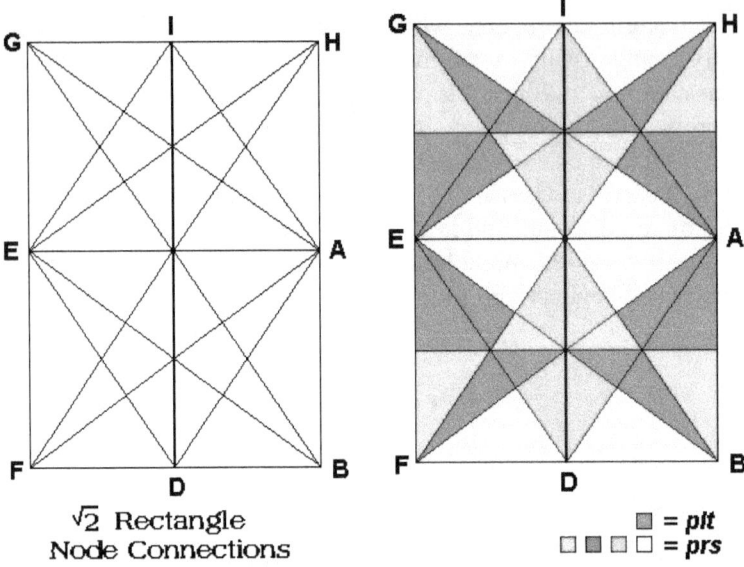

√2 Rectangle
Node Connections

☐ = plt
☐ ☐ ☐ ☐ = prs

Appendix Diagram 7a (left) and 7b (right)

For clarity we have left out the lines between nodes GD, IF, IB, DH. The large triangles formed between these nodes will be plt triangles.

The uniqueness of the sqrt2 rectangle shows up dramatically in Diagrams 7a-7b. The sqrt2 rectangle is truly a unique geometric figure, not only in its replicative geometry but also in its direct relationship to the tetrahedron. Because it is the only rectangle whose regular division produces smaller rectangles of the same proportions, all the triangles formed within the rectangle are repetitions of the geometry found in the two larger triangles from which the original rectangle is formed (these were prs triangles). Calculations have shown that only the sqrt2 rectangle has as many repetitions of its formative right triangle within lines drawn between the nodes -- over twice as many repetitions as in other rectangles. Furthermore, the triangles formed within the rectangle repeat the characteristic internal geometry of the tetrahedron cross-section.

This being the case, we turn our attention to the cross-section as it appears in the rectangle. Diagram 8 below shows the two interlocked cross-sections in the bottom half of the rectangle. If two tetrahedra were superimposed one on the other so that they share a common axis, their superimposed cross-sections would match this diagram (common axis is EB)

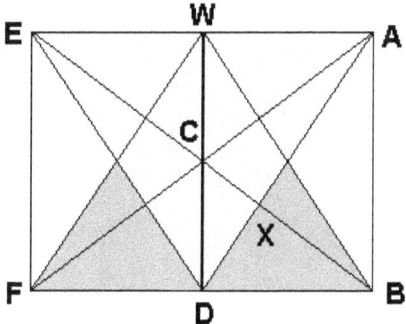

Appendix Diagram 8

Tetrahedron Cross-Sections EAD and BFW within a sqrt2 rectangle.

Point C is the center of both tetrahedra. Of course, EABF is a sqrt2 rectangle (lying on its side), since it is in effect the lower 1/2 of the rectangles shown in Diagrams 7a and 7b.

With this figure we arrive at an unexpected relationship between the sqrt2 rectangle and the tetrahedral latitude. It turns out that when a circle is drawn with its center at C and its radius the length CB, this circle will represent a cross-section of a sphere enclosing the intelaced tetrahedra having cross-sections EAD and BFW. The center of the tetrahedron C is also the center of the enclosing sphere. Line EB represents the axis of both sphere and tetrahedron, and line V-V' is the equator of the sphere.

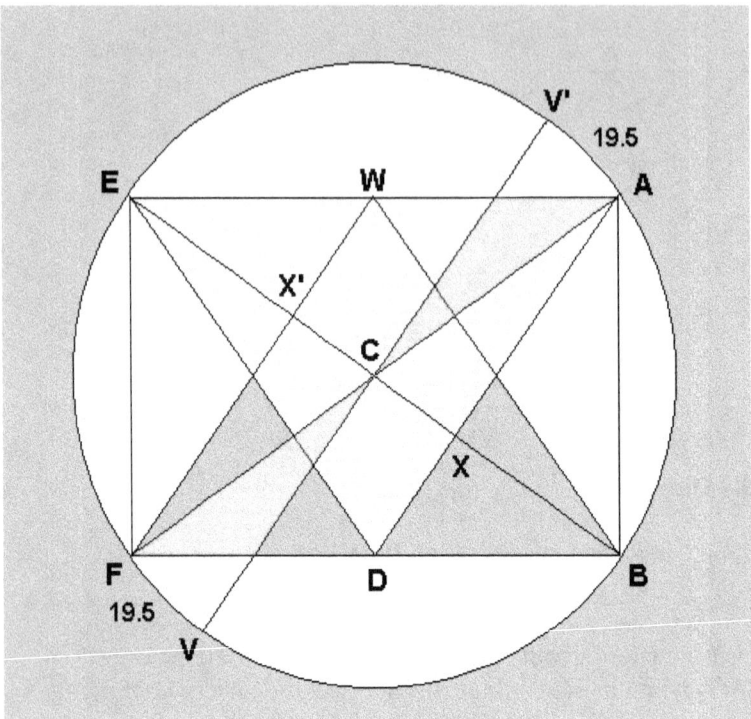

Appendix Diagram 9

The sqrt2 rectangle and interlocked tetrahedron cross-sections within the cross-section of a surrounding sphere.

This analysis shows that the sqrt2 rectangle contains all the information necessary to identify the position of the 19.5 degree tetrahedral latitude and to draw an accurate cross-section of a sphere enclosing two interlocked tetrahedra. A surprising result of this analysis is that a sqrt2 rectangle inserted within a sphere so that two diagonally opposite corners of the rectangle are located at the poles, the other two corners of the rectangle will touch the sphere at the tetrahedral latitude points -- one North and one South. Thus the tetrahedral latitude can be determined by the sqrt2 rectangle without direct reference to the tetrahedron.

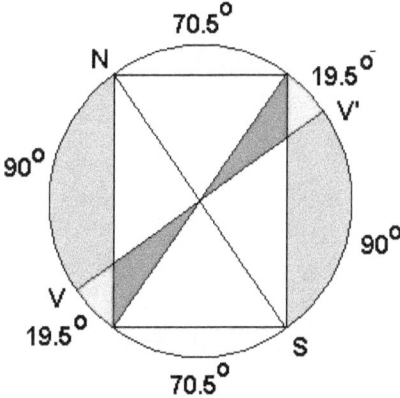

Appendix Diagram 10

The sqrt2 rectangle alone defines the tetrahedral latitude on an encircling sphere. (N = North Pole, S= South Pole, V-V' = equator.)

In Diagram 10, yet another surprise: the arcs of the circle, which are defined by the corners of the sqrt2 rectangle and the equator, occupy 90, 70.5, and 19.5 degrees -- just the angles found in the plt triangle which was represented many times in the tetrahedral cross-section. This graphically demonstrates the close correspondence between the geometry of the sqrt2 rectangle and the geometry of the tetrahedron. In its internal representation of one of the five regular solids (the tetrahedron), the sqrt 2 rectangle is a geometrically "rich" figure.

7. The Mound Pentad and the Rectangle

Finally, we come to the connection between the mound pentad and the sqrt2 rectangle. Diagram 11 shows the area of the rectangle outlined by the five mounds GABDE. As you can see, the tetrahedral cross-section is included in the pentad (EAD, axis EX).

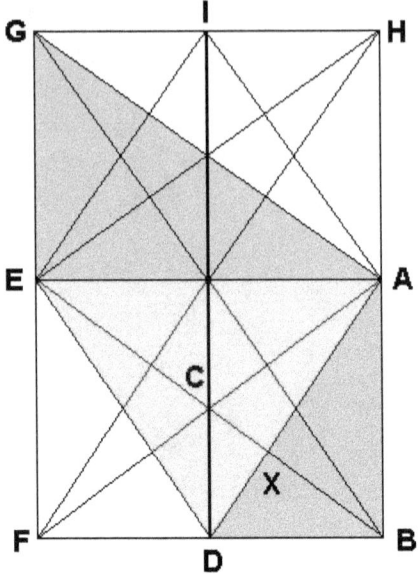

Appendix Diagram 11

The area outlined by mounds GABDE within the sqrt2 rectangle.

It is true that only five of the eight nodes of the sqrt2 rectangle are represented by mounds. Certainly as far as we can detect, there are no mounds visible at the remaining three nodes. However, five mounds in the configuration shown are just the minimum necessary to make a strong geometric reference to the sqrt2 rectangle and the cross-section of the tetrahedron.

Mounds EAD form the cross-section; the line between mounds EB marks the axis; and mound G not only introduces a large prs triangle from which the rectangle may be inferred but also supplies the necessary opposite corner of the rectangle. Note that it was the presence of mound G, and the consequence that within even the tetrad of mounds GADE there are two right triangles and one isosceles, that led Dr. Crater to discover the geometry that most closely fits the mound formation. But only with the addition of the fifth mound, B, is the inference to the sqrt2 rectangle obvious (because G and B form two corners of the rectangle).

From the point of view of economy of means, the pentad configuration is a highly efficient reference to the sqrt2 rectangle and the tetrahedral geometry that it contains.

8. Closeness of Fit

It is important to understand how closely the mound configuration actually fits the ideal pentad (which as shown above is a large portion of the sqrt2 rectangle). Dr. Crater has calculated that the average variation from the estimated mound centers is only 0.94 pixels or slightly less than 150 feet. In contrast the distances between the pentad mounds are about two miles. Furthermore the mounds are not large, being on average only about 8 x 8 pixels. Within the error margin for the mound centers the nodes of the ideal figure are on the centers of the mounds. The chart below refers to the angles between Cydonia mounds GADEB forming the pentad of mounds. Angles are in degrees. Each triangle among mounds GADEB is represented by three letters. Triangle ADE is the "lrr" tetrahedral cross-section as explained above. Note the four similar right triangles AGD, EAG, BAD, and AEB. These are "prs" right triangles.

Since there are three angles for each triangle in the mound configuration and also in the ideal pentad, there are six lines altogether. However, the ideal and measured angles are so close that there appear to be effectively just three lines. The close overlap of the lines for the ideal angles and the lines representing the measured angles indicates the closeness of fit between the ideal and measured figures. Within measurement error, the mounds are clearly distributed according to the ideal configuration. According to Dr. Crater's calculations, the chances of this close fit occurring at random are less than one in two hundred million (based on computer simulations of random distribution patterns).

9. Conclusion

At the outset we outlined several questions appropriate for evaluating the possibility of artificiality. One of those questions was "Is there any recognizable cultural or symbolic quality in the patterns of mounds that might provide a clue to possible meaning or utility?" There are in fact some known terrestrial cultural contexts in which the square root 2 (sqrt2) rectangle has a role. In his book The Geometry of Art and Life, Matila Ghyka discusses "Greek and Gothic Canons of Proportion" in which it is said that the proportions of certain rectangles, derivable geometrically from the square, represent harmonic balance in art and architecture. Footnote: These include the sqrt2, sqrt3, and sqrt5 rectangles. They are referred to as "dynamic" rectangles. According to Ghyka, These "dynamic" rectangles were thought to produce "the most varied and satisfactory harmonic subdivisions and combinations" for use in art and architecture. In his book, he shows seven different "harmonic decompositions" of the sqrt2 rectangle, which would be used to establish a great variety of proportions perceived as aesthetically pleasing and as a basis for symbolism in art and architecture.

Among traditional classic proportions having a prominent place in design, the "golden section" or "golden ratio" (1:1.618) is perhaps much better known than the various tetrahedral proportions found in the sqrt2 rectangle. When researching the geometry of the rectangle, repeated efforts were made to determine whether the various proportions and ratios generated within the rectangle would yield, in addition, the golden section. These efforts proved unsatisfactory. However, recently I discovered a paper of Mark A. Reynolds in which a most elegant geometric solution to the problem is presented.

The problem tackled by Reynolds was to discover the golden section of the circumference of a circle (what we might call the "golden arc"). Since the circumference of a circle =360 degrees, the ratio 1 to 1.618 is derived to be 137.5 to 222.5 degrees (rounded to one decimal).

Footnote: To construct a sqrt2 rectangle from a square, one need only start with the square and scribe the arc of a circle having as its center one corner of the square and its radius the diagonal of the square. Where the arc meets an extended side of the square determines the length of the long side of the resulting sqrt2 rectangle. This arc (having radius AC) is visible on the right in Diagram 12 below. Where the arc encounters an extension of side AD establishes point E, in the ratio of sqrt2 to side AB. Thus ABEF is a sqrt2 rectangle.

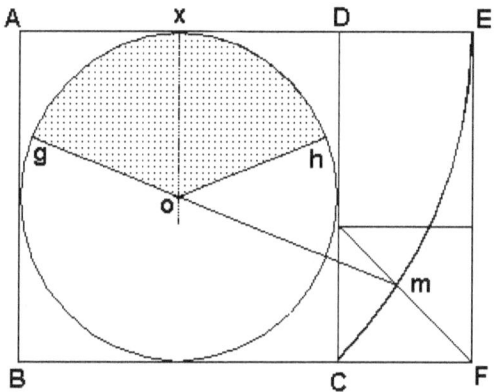

Appendix Diagram 12

Mark Reynolds' derivation of the golden arc from the sqrt2 rectangle

Having drawn the arc, a circle is placed in the initial square with a diameter equal to the sides of the square. The remainder of the sqrt2 rectangle is then divided by constructing a square of sides equal to length CF (the small rectangle above this square is again a sqrt2 rectangle).

Point m is now determined as the intersection of the arc with the diagonal of the small square. When a line is drawn from m through the center o of the large circle, angle gox = 68.75 degrees within a small margin of error. Doubled, this yields angle goh, the desired golden ratio arc of 137.5 degrees (the remaining arc being 222.5 degrees.) In addition to this particular derivation,

Reynolds found three other pathways within the sqrt2 rectangle that yielded similar results. Thus not only are the "dynamic" architectural and aesthetic values sqrt2, sqrt3, sqrt5 present in the sqrt2 rectangle, along with the complete representation of the geometry of the tetrahedron, but also the classic golden ratio can be derived from the rectangle by simple geometric means.

One of the probable uses of such geometric measurement in the distant past lies in the possibility of laying out architectural lines having symbolically significant ratios, e.g. on the ground of a proposed site by pacing out first a square, drawing an arc on the radius of the diagonal to create a sqrt2 rectangle, and then inferring further values by utilizing the productive elements of the rectangle (such as the golden ratio as described).

In a series of two articles John A. R. Legon presents data indicating that the layout of the three pyramids at Gizeh, Egypt (including the Great Pyramid) is based on a rectangle having as its sides sqrt 2 and sqrt 3, with its diagonal being sqrt 5.

Footnote : The values sqrt2 and sqrt3 are predominant in the sqrt2 rectangle (as well as sqrt6), and the value sqrt5 is derivable geometrically from that rectangle. From Legon's work it would appear that the layout of the Gizeh pyramids may have been influenced by the same concepts of harmonic proportion as those discussed by Ghyka.

What is most interesting about the Legon data is that it implies an application of the "dynamic rectangle" concept to the distribution of a group of architectural structures. On Mars we have, perhaps, an analogous situation. At Cydonia there is a group of formations of relatively uniform size (the mounds), distributed according to the "dynamic" sqrt 2 rectangle. The cultural implication may be that the distribution of mounds (if they are artificial) is architectural or aesthetic in intent. If so, this would be a piece of information regarding the cultural mind-set of the builders. And since the Canons of Proportion are geometrically derived -- geometry being a universal science -- there is no reason to suppose that extraterrestrial intelligence might not be responsive to the same concepts of harmonic proportion as those appreciated in the terrestrial cultural tradition.

Another possibility is that the distribution of mounds had a communicative intent, that is, it may have been intended as a kind of signal. Some critics have said that any "message" from extraterrestrials ought to be simple and easy to recognize; that any intentional pattern would have to be "obvious." But this is not necessarily the case. It could well be that a signal was created that could only be understood by a civilization advanced enough in mathematics and geometry to interpret and respond to a more complex geometric pattern. Or it could be that the "message," if that is what it is, speaks to a symbolic and aesthetic side of culture rather than a purely scientific one.

Of course, the pattern of mounds may be simply a natural formation; but if it is, it would appear to be a geological oddity because of the very low probability of the distribution having occurred by random forces. Against the idea of a natural origin, we have developed answers to our initial questions that would seem to suggest otherwise:

The mound geometry conforms to a rectangular grid.
The geometry is not commonplace, but unique.
The geometry is not unproductive, but rich.
The geometry conforms to a recognizable mode of architectural aesthetics and is capable of expressing symbolic meaning.

We conclude that the probability of either possible intentional design or radical geological anomaly for the mound configuration at Cydonia is more than sufficient to warrant high priority for continuing active investigation of this site.

**********   *********   ***********   **********   *******

[Appendix Figure 78 e]

The author with professor Stanley McDaniel in San Francisco, California, April 1994

Chapter 6

PREVIOUS PUBLICATIONS ABOUT ANOMALIES ON MARS

[Figure 79]

NASA sent two spacecraft which arrived at Mars in 1976. Viking 1 and Viking 2 remained in orbit and took the original photos which comprise the early part of the analysis detailed in this book. Both Vikings also put down landers, each of which carried portable laboratories designed to conduct experiments designed to search for "life" - meaning biological organisms such as bacteria in the soil.

The booklet shown above, NASA SP-408, was published in 1976, expounding on some of the early results from the Viking orbiters and landers. The final chapter in the above book was entitled, "The Search

For Life". This chapter says, "It can be assumed that living organisms are reasonably well adapted to their environments and that they are composed of chemicals that are available to them. In any case, knowledge of the organic and inorganic chemicals found in the surface materials and the atmosphere and of the physical state in which they occur, provides boundaries on the kinds of biochemical reactions that might be detected."

Naturally, the designers of the three 'life' experiments had to make some assumptions prior to the landers sojourn on the surface of Mars. In a paradox, the Viking landers made discoveries that could have refined the design of the initial experiments which had taken place on Earth many years before the successful orbits and subsequent landing of the two Viking spacecraft.

NASA SP-408 states:

"As it turned out, the results have been at once surprising, puzzling and scientifically stimulating." (page 59)

The three experiments were:

(1) the pyrolitic release (PR) experiment which looked for the biological synthesis of organic molecules

(2) the labelled release (LR) experiment which searched for the assimilation of labelled nutrients with subsequent release of gases

(3) the gas exchange (GEX) experiment which looked for metabolically caused changes in the composition of any gases in contact with living organisms.

Although all three experiments gave an overall positive signal for "life" NASA has spent the last forty years or more denying the results and **explaining them away in terms of chemical reactions**. The contention about the Viking lander experimental results continue to this day and is an indication of the fact that the scientific community on Earth is rather unprepared for the discovery of life on other planets - even just microbial life. But of course, if one believes in the modern scientific theory of evolution, then one has to admit that microbes can and probably evolve into more complex forms of life. A Catch-22 situation for the sceptics! (See the letter in "Scientific American", 10 October 2019 by Dr. Gilbert Levin, designer of the "Labelled Release" experiment.)

NASA's 1976 publication stated that **"the results of the LR experiment were startling initially and continue to present surprises."** (page 62)

With regard to the GEX, it states: **"The results to date of this experiment have also been surprising."** (page 63)

Describing the PR (pyrolitic release) it indicated: **"This is a numerically significant peak"** about the measured radiation level. The net level of the second peak was 96 counts per minute. The predicted level in the absence of any carbon assimilation was only 15 counts per minute.

*** **** ***** ******

LANDING SITES

A US rover called Perseverance will land in Jezero Crater, near a delta formed by an ancient river — a prime location for finding signs of past life if it existed. China is considering several landing sites for its Tianwen-1 mission.

Tianwen-1 potential landing sites   • Previous missions

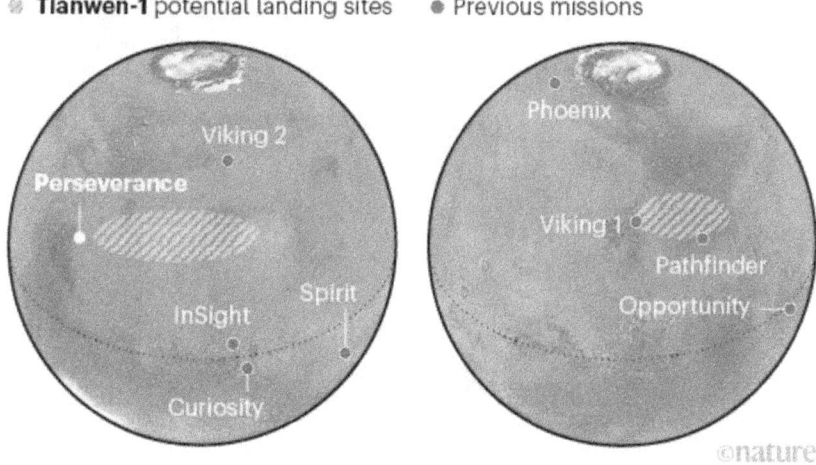

Landing sites, compiled by E. Lakdawalla/Planetary Society; Mars base map, US Geological Survey.

[Figure 79 a - Courtesy Nature]

Viking 1 and 2 lander sites in relation to later missions: Curiosity, Spirit, Insight, Pathfinder, Phoenix and Opportunity.

The 2020 launch of Perseverance will land the latest NASA rover in Jezero crater in 2021.

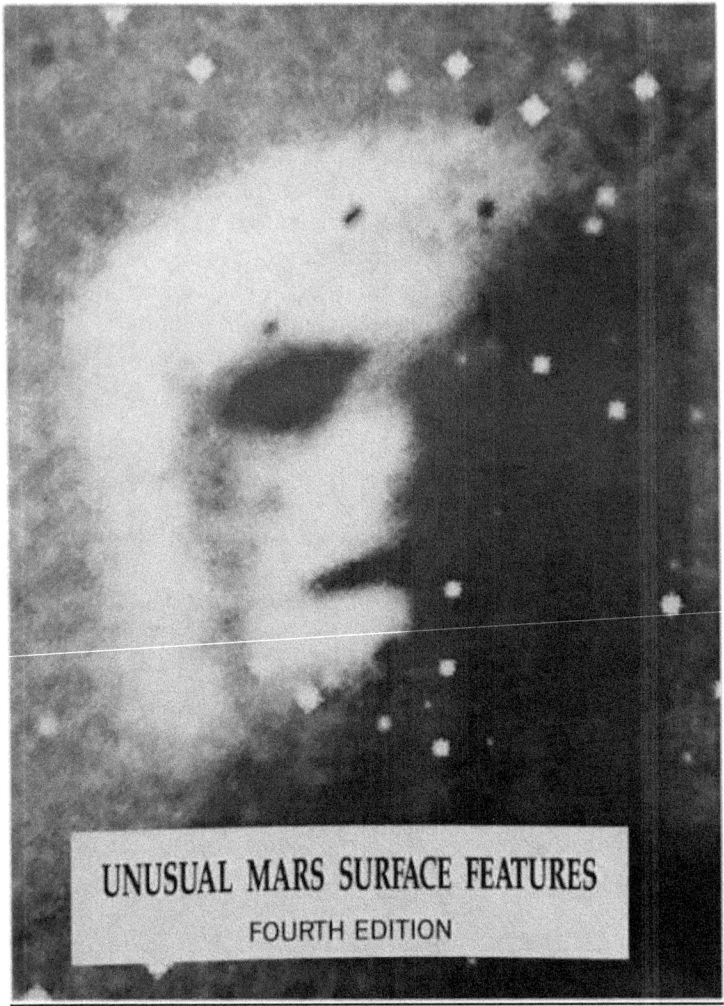

[Figure 80]

The first edition of **"Unusual Mars Surface Features" (UMSF)**, written by Vincent DiPietro and Gregory Molenaar was published in 1982. The fourth edition in 1988 had contributions from Dr. John Brandenburg. This book brought to the fore the famous "Face On Mars" seen in Cydonia and clearly visible in Viking 1 frames 35A72 and 70A13, as well as some pyramidal looking hills in the area.

The photographs published in UMSF had been enhanced by a process known as "Starburst Pixel Interleaving Technique", extracting more detail from the 47 metres per pixel Viking 1 images.

The controversy about the "subjective" impression of a massif looking like a human-type face continued for many years until NASA released a poorly processed photo, taken in April 1998 by the Mars Global Surveyor.

This contrast enhanced image, (referred to by some researchers as "The Catbox") modified with an extreme high-pass filter, removed tonal variations, giving the impression that the "Face On Mars" is a flat, featureless formation. That release probably provided the "kiss-of-death" to the discussions about whether this mile-and-a-half rock was a representation of a face or a naturally eroded landform, or a natural massif modified by intelligence.

However, NASA's MRO has taken later photos of "The Face" which indicate the feature still retains some perplexing anomalies. These are surprisingly similar to what was first observed in 1976 in Frames 35A72 and 70A13.

[Figure 80 a]

Vince DiPietro and Gregory Molenaar, authors of "Unusual Mars Surface Features".

The analysis reported here and now, performed in an exhaustive manner by Crater and McDaniel, was quite independent of the facial feature in Cydonia. The mathematics of the mound layout stands by itself as an objective, non-refutable signpost into the future exploratory travels in the vast arena of the solar system - by the next generations of space travellers from many different nation around our globe.

[Figure 81]

The book "The Face On Mars - Evidence for a Lost Civilization", written by Dr. Randolfo Rafael Pozos" was published in 1986.

[Figure 81 a]

Professor Stanley McDaniel with this author and Dr. Randolfo Pozos.

[Figure 81 b]

This author with Dr. Randolfo Pozos - April 1994, San Francisco

[Figure 81 c]

Dr. Randolfo Pozos discussing this author's findings on Frame 70A11 to the west of the features already identified in Cydonia.

This finding of a secondary face and an area dubbed "The Sirisena Quadrangle" by professor McDaniel, showed a "Star Pattern" and a circular entrance, which was dubbed by this author as "Crater's Cave" in honour of Dr. Horace Crater.

Later images taken by NASA confirmed the existence of the features but their interpretation remains open.

[Figure 82]

The first edition of Hoagland's book was published in 1987. Later editions were issued in 1992 and 1996.

[Figure 82 a]

Richard C. Hoagland with TV comedian Jay Leno at the Los Angeles UFO Expo in 1994. Jay Leno is holding a model of "The Face On Mars" made by sculptor Kynthia. This photo was supplied by David Laverty, as was the photo below of Jay Leno with Richard Hoagland.

[Figure 82 b]

Hoagland is a science writer and broadcaster.

[Figure 83]

"The Martian Enigmas - A Closer Look" was written and published by Dr. Mark Carlotto in 1991.

[Figure 83 a - The author with Robert Johnston and Dr. Mark Carlotto in the UK]

[Figure 84]

Dr. Mark Carlotto earned his Ph.D. in Electrical Engineering from Carnegie-Mellon University in 1981.

From 1981 to 1983 he was an Assistant Adjunct Professor at Boston University. Dr. Mark Carlotto provided an excellent set of computer enhanced images of "The Face" and "D&M Pyramid" in Cydonia. The D&M Pyramid is named after DiPietro and Molenaar.

The picture above shows the back cover of Carlotto's book. The image on the left shows the face in profile view, as would be seen from the surface of Mars, looking east.. This enhancement was done by Dr. Mark Carlotto as part of his early investigation into the features in Cydonia.

(ESA, in 2003, also supplied a profile view of the face, as would be seen from a short distance away.)

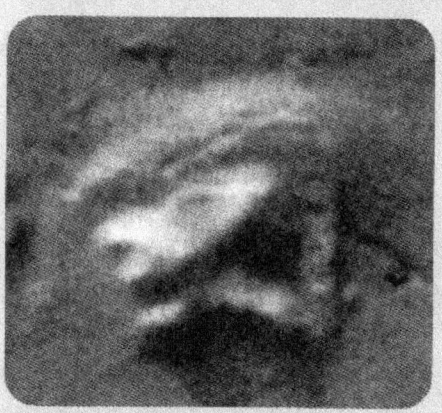

[Figure 85]

"The McDaniel Report" was written by emeritus professor Stanley V. McDaniel and published in 1993.

[Figure 85 a]

Professor Stanley V. McDaniel studying Viking images with Dr. Randolfo Pozos. Photos taken by the author in San Francisco in April 1994.

[Figure 85 b]

The McDaniel Report

In preparation before the reported loss of the Mars Observer spacecraft on August 21, 1993, this thorough and detailed year-long study is an account of NASA's apparent failure to fulfil its legal, scientific, and ethical responsibilities to the American people by properly evaluating data researched by highly qualified independent investigators, and by assigning mission priorities accordingly. What this research implies could lead to an economic and scientific expansion of unprecedented proportions in the United States and other spacefaring nations. Yet against the background of this significant positive potential, NASA has been regularly giving members of Congress and the public false or misleading statements, leaving government leaders misinformed and unaware of the full story. In the near future, decisions will be made in Washington regarding the future of the space program. This study is a comprehensive briefing on the Martian potential for any government official or member of Congress involved in such decisions. The information in this report is highly relevant to future NASA missions, missions by other countries, and to the Mars Observer if communication with that craft is restored.

What Scientists Say about the McDaniel Report

- "This important document makes a well-documented, quote-by-quote and letter-by-letter case that NASA chose scientifically inappropriate responses to the issue of Martian anomalies in the Cydonia region and the testing of the hypothesis that they might be artifacts...The impact of verification of the hypothesis is so great that no simple, unique opportunity to test such ideas [e.g. high-resolution photography] should be overlooked. Testable, falsifiable, scientific hypotheses should be treated as such. Ridicule (which has been used by NASA spokespersons in connection with the Martian anomalies) should always be out of bounds in a scientific context."—*Dr. Thomas Van Flandern, former Head, Celestial Mechanics Branch, U.S. Naval Observatory*

- "The analysis that DiPietro, Molenaar, Hoagland, Carlotto, Torun, and others have done on the two Viking Photographs taken of the Cydonia Region of Mars surrounding the Face is very intriguing and thorough. The claims based on this analysis are that these features are unlikely to be entirely due to natural erosive forces. This is an understatement of extraordinary magnitude."—*Dr. Horace Crater, Professor of Physics, The University of Tennessee Space Institute*

- "The so-called "Face on Mars" is unlike any natural feature I have ever seen or heard about. To ascribe this feature of such symmetry and uniqueness to 'wind erosion' is to plead a special case for a geologic process with no supporting evidence."—*James Berkland, Certified Engineering Geologist, Former geologist for the U.S. Geological Survey and the U.S. Bureau of Reclamation*

- "I am convinced that it is not only worthwhile, but extremely important, that the investigations into the Artificiality Hypothesis be continued utilizing the best techniques and resources available among the scientific community."—*Dr. Robert M. Schoch, Associate Professor of Science and Mathematics, College of General Studies, Boston University*

ISBN 1-55643-088-4

North Atlantic Books

[figure 86]

[Figure 87]

The second edition of Dr. Mark Carlotto's book was published in 1997. Chapter 6 in this book discussed "Other Intriguing Objects On Mars" including "The Crater Pyramid" from Viking Frame 43A01, "The Runway" from Viking Frame 86A08, "The NK Pyramid" from Viking Frames 70A09 and 70A11. Also included in this chapter were two anomalous objects found by this author on Frame 70A10 and dubbed "Fort Aetherius" and "The King Pyramid".

Dr. Mark Carlotto's superb 3-D rendering of the Cydonia plain makes a striking cover to this book.

[Figure 88]

The controversy about the Viking lander experiment results has been thoroughly documented in the book: **"MARS - The Living Planet"**, written by Barry E. DiGregorio with Dr. Gilbert V. Levin and Dr. Patricia Ann Straat.

This book published in 1997 is highly recommended for those interested in the finer details of the experiments and the ignominious attempt by NASA to explain away the results. Such action by NASA makes a mockery of the fact that the scientific method is said to be "empirical".

Since 1976, NASA has not sent a single experiment to Mars to search again for microbial life amidst its sands. The 2020 mission will do so.

DiGregorio asks in his book: **"Mars - The Living Planet"**

"With new evidence that life existed on Mars billions of years ago, there is no scientific reason why that life may not have continued to evolve right up until today..
"Yet, it is repeated year after year that "the Viking Lander Biology Experiments searched but did not find any evidence of life on Mars", as the quote reads that is emblazoned on the Viking exhibit at the Smithsonian Institute.
"That statement is both unscientific and untrue.. **The fact that this claim has been perpetuated by NASA for more than twenty years despite evidence to the contrary is nothing less than suspicious.** There is only one way to be certain about life on Mars, and that is to return to Mars with updated versions of life-detection experiments designed to settle that issue once and for all. Since Viking, NASA has turned its back on settling this incredibly important issue. Why?" - Barry DiGregorio, 1997.

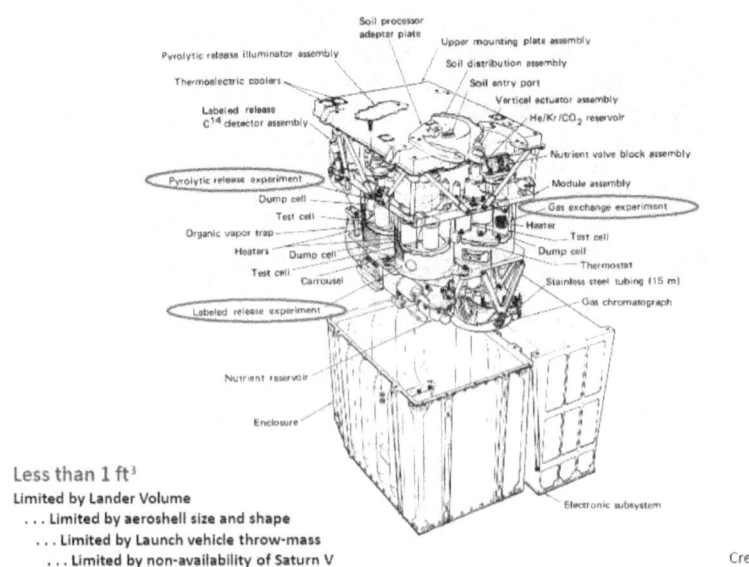

The biological package carried by the Viking 1 and Viking 2 landers to search for evidence of life. (Image credit: NASA)

[Figure 88 a]

NASA's twin Viking landers touched down on Mars in 1976 to hunt for signs of life on the Red Planet. Forty years later, scientists are still arguing about what the landers' observations mean. (Image: © NASA)

[Figure 88 b]

[Figure 88 c]

Professor Stanley McDaniel together with the author and Daniel Drasin discussing the Viking Lander life experiment results. Photo taken at the "Goro Robata" restaurant in San Francisco, California on 17th April 1994. Daniel Drasin was one of the early researchers.

[Figure 88 d]

The author with professor Stanley McDaniel and Daniel Drasin at the Whole Life Expo in San Francisco, California, April 1994, where we ran a workshop together on the discoveries in Cydonia.

[Figure 88 e]

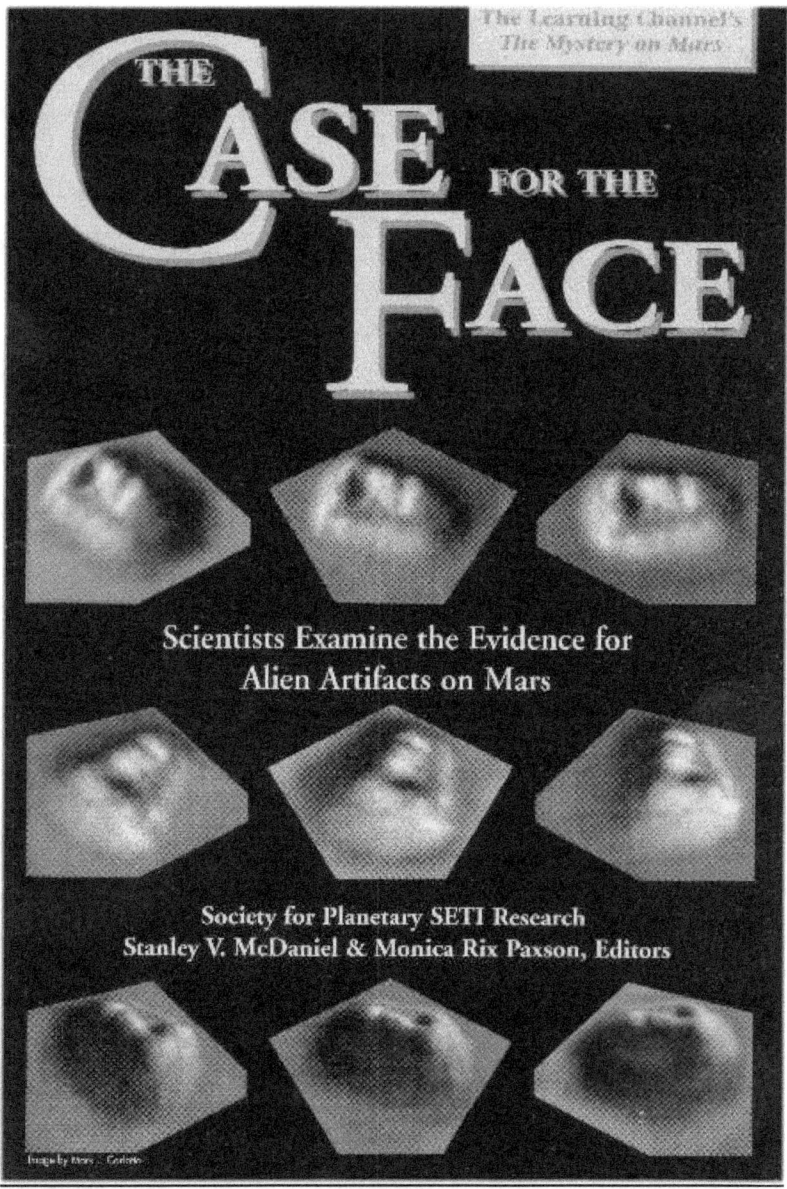

[Figure 89]

Written and published by SPSR members in 1998.

http://spsr.nmcc.edu

[Figure 90]

Back cover of the 1998 book by SPSR. Dr. David Webb stated: "The best scientific evidence ever assembled for artifacts of advanced alien life."

Arthur C. Clarke said, "I am more than willing to accept the possibility of alien artifacts in the solar system - after all, that's what 2001 was all about! So I welcome continued interest in the Cydonia Mystery and deplore attempts to dismiss it."

"Dead Mars - Dying Earth" published in 1999. Written by Dr. John Brandenburg together with Monica Paxson

[Figure 90 a]

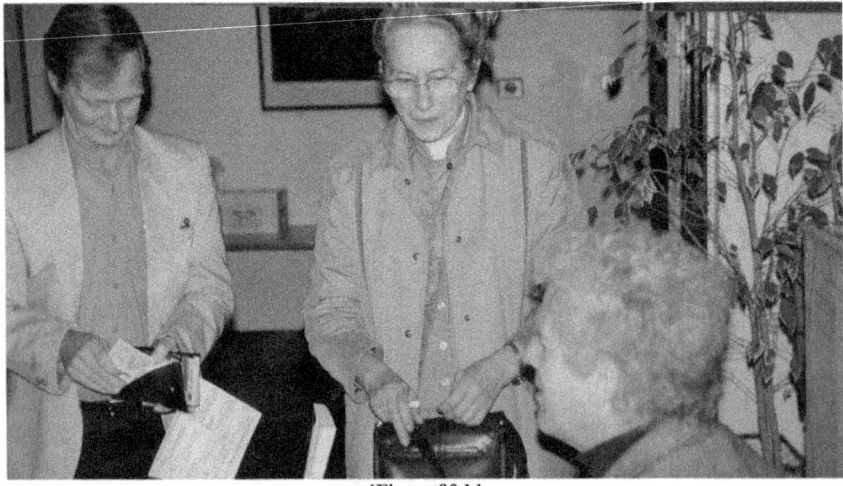

[Figure 90 b]

(Photos by Ananda Sirisena)

Dr. Brian O'Leary, a former Apollo scientist-astronaut and Princeton University Physics Department lecturer gave a talk and demonstration at Aylesbury Grammar School in England on 23rd November 1997. The pictures above show O'Leary discussing his books: "Exploring Inner And Outer Space" and "Miracle In The Void". Brian O'Leary was a member of SPSR until his demise in South America.
He was also the author of: "The Second Coming Of Science".

[Figure 90 c]

Malcolm Smith, Ananda Sirisena, Greg Fewer and professor Dr. Horace Crater outside the British Interplanetary Society in London - 2006.

[Figure 90 d]

Dr. Mark Carlotto, professor Horace Crater and James Erjavec at a meeting in Boston. (Picture by Mitchell Swartz)

[Figure 90 e]

Professor Stanley McDaniel in conversation with Daniel Drasin at the Whole Life Expo in San Francisco, California in April 1994.
(Photo by Ananda Sirisena)

[Figure 90 f]

The author with Rhonda and Kynthia in San Francisco, April 1994. Kynthia sculpted an accurate model of "The Face On Mars" from the Viking data - its shadows providing the height information.

[Figure 91]

Chris O'Kane who worked with a group of school children in Scotland and studied "The NK Pyramid" on Mars published an E-book titled "Journey to Cydonia: The First Small Step" in 2018.

[Figure 91 a]

Chris O'Kane, author of "Journey To Cydonia: The First Small Step" at his 27th June 2020 presentation on the Martian Revelation radio show.

EUROPEAN

PAPER

SIZES

The dimensions of the A series paper sizes, as defined by the ISO 216 standard, are given in the table to the right of the diagram in both millimetres and inches (cm measurements can be obtained by dividing mm value by 10). The A Series paper size chart, below left, gives a visual representation of how the sizes relate to each other - for example A5 is half of A4 size paper and A2 is half of A1 size paper.

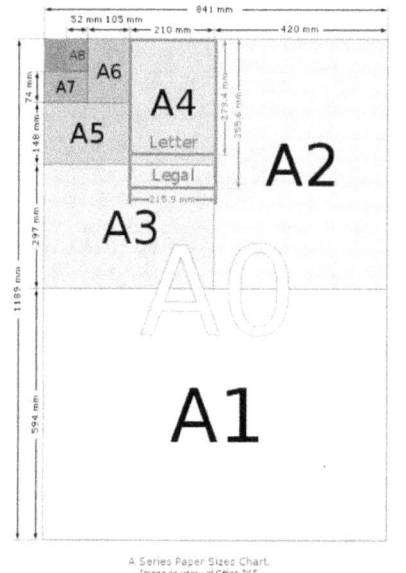

Table of Paper Sizes From 4A0 to A10

Size	Width x Height (mm)	Width x Height (in)
4A0	1682 x 2378 mm	66.2 x 93.6 in
2A0	1189 x 1682 mm	46.8 x 66.2 in
A0	841 x 1189 mm	33.1 x 46.8 in
A1	594 x 841 mm	23.4 x 33.1 in
A2	420 x 594 mm	16.5 x 23.4 in
A3	297 x 420 mm	11.7 x 16.5 in
A4	210 x 297 mm	8.3 x 11.7 in
A5	148 x 210 mm	5.8 x 8.3 in
A6	105 x 148 mm	4.1 x 5.8 in
A7	74 x 105 mm	2.9 x 4.1 in
A8	52 x 74 mm	2.0 x 2.9 in
A9	37 x 52 mm	1.5 x 2.0 in
A10	26 x 37 mm	1.0 x 1.5 in

A4 IS HALF THE SIZE OF A3 PAPER
A4 IS LARGER THAN SIZE " LETTER"
A4 IS SMALLER THAN SIZE " LEGAL"

[Figure 92]

A search on the internet provided the above table of paper sizes. Why is this table of interest to us when trying to understand and explain this unresolved mystery on Mars?

The reason is that European paper sizes are based on a square-root 2 grid! If one measures the A4 size paper to be 210 mm by 297 mm, it is evident that the short edge of the paper, if it were one unit would mean that the long edge of the sheet of A4 paper would be root2. If 210 mm = 1 unit, 297 mm = root 2.

If one folds the A4 sheet of paper in half, one gets the A5 paper size, which also has ratio of the sides 1: root 2. If one folds an A3 sheet in half one will end up with an A4 sheet. Thus, the A3 also has edges in the ratio 1: root 2. This makes the European paper sizes self-referential.

As a simple exercise, take an A4 sheet of paper (210 mm x 297 mm) and fold it four times : A4 -> A5 -> A6 -> A7.

The resulting A7 sheet of paper will measure 74 x 105 mm. Now open up the sheet and along the creases one will be able to place marks at the fold edges and end up with the pentad and hexad pattern!

One cannot do this with American paper sizes, nor with the older British paper size known as "foolscap". Try it for yourself.

The pattern you have just created with the A4 sheet of paper matches the mound configuration layout at a much smaller scale. The 1:root 2 ratio was picked for its aesthetic and self-referential facets - it did not happen by chance. There was human intelligence and choice behind the selection of European paper sizes, arising from an older DIN standard.

Therefore we could conclude that the mounds on the Cydonia plain were not randomly placed but were part of an intelligent architectural layout. The originating intelligence understood the internal geometry of a regular tetrahedron, **which most residents on Earth are not aware of.**

CHAPTER 7

A WALK IN THE WOODS OF CYDONIA

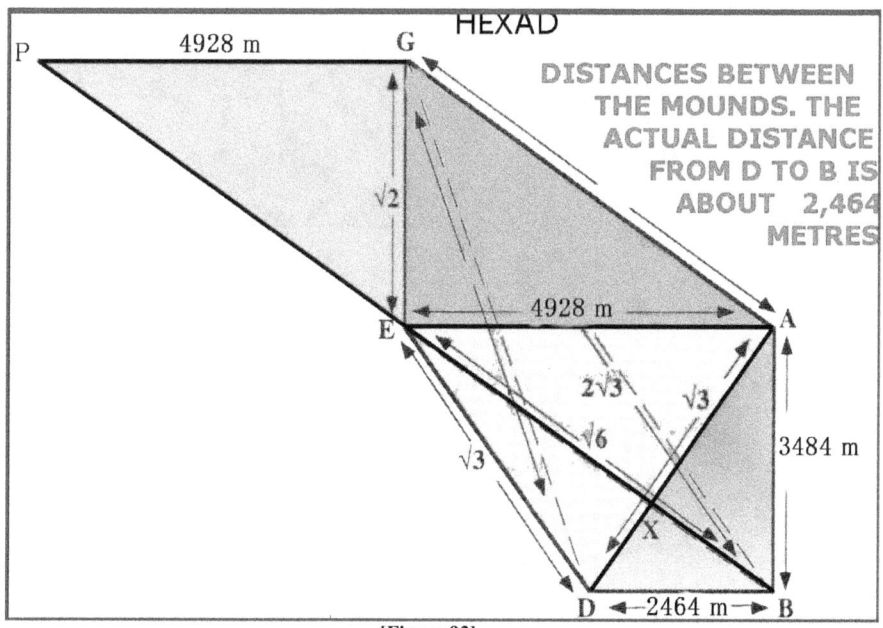

[Figure 93]

Imagine that there was a copse in Cydonia a very long time ago. One could have walked from mound to mound, from D to B and then A, assuming atmospheric conditions were similar to Earth today and the planet Mars was able to sustain life as we know it now.

It would be a wonderful walk in the woods....

The diagram above shows the distances between all the six mounds of the hexad in metres. The basic unit distance DB is approximately 2,464 metres.

Distance from G to D = 7392 metres.

In kilometres, the total distance around the perimeter is 27.2.

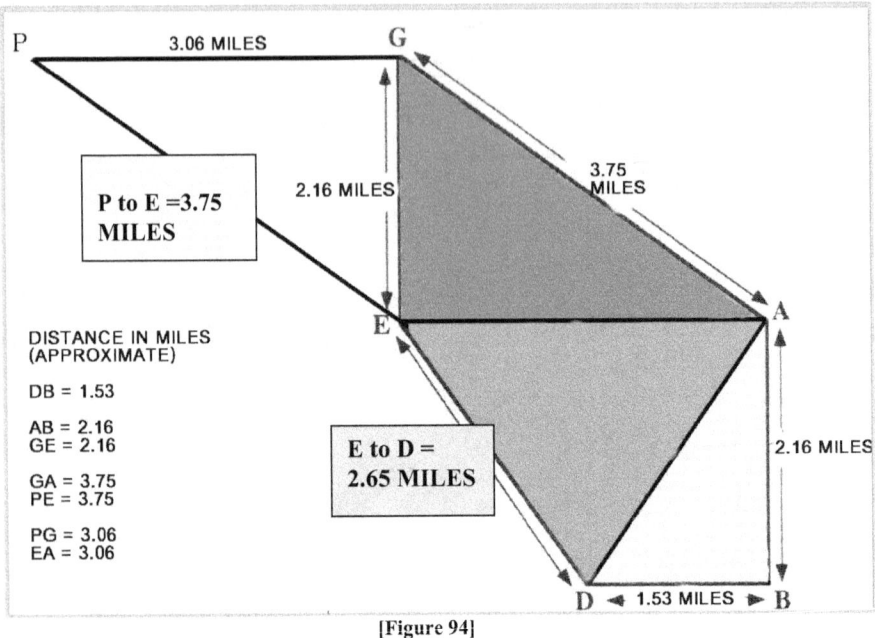

[Figure 94]

The same layout as in the previous image, here showing the distances between the mounds in miles. These are easily walkable, or runnable, by a fit individual on Earth. Perhaps there were fit living beings on Mars a long time ago?

The distance from D to E is 2.65 miles approximately.

A walk around the entire perimeter of the hexad would be about 16.9 miles.

Professors Crater and McDaniel had written in their paper:

"The hexad shows a six-mound pattern with a co-ordinated fit involving similar right triangles and a related isosceles triangle. The stringent requirement for a co-ordinated fit severely restricts the number of **chance matches**."

APPENDIX A

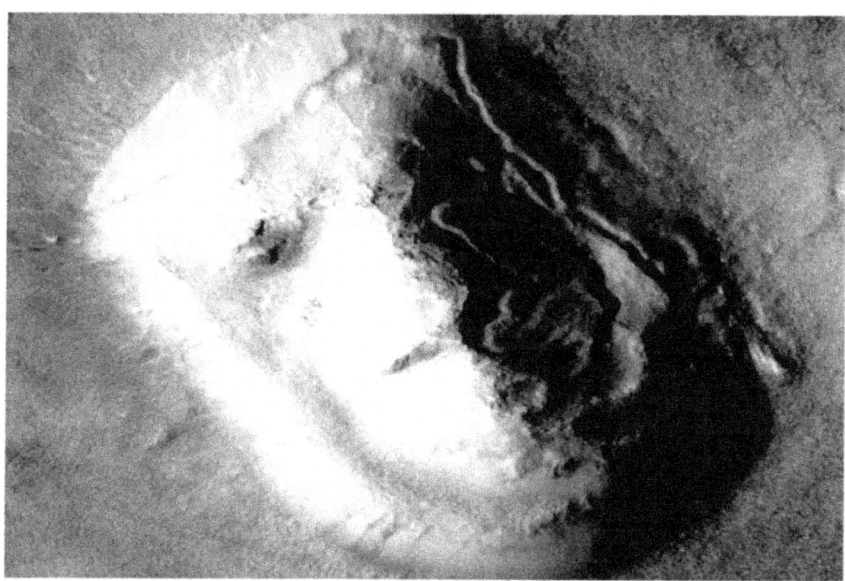

[Figure 95]

The "Face On Mars" photographed in 2009 and quietly placed on a NASA website. This high resolution image confirms predictions made in the 1980s and 1990s of 'teeth in the mouth' and eye details by experts in digital enhancements.

[Figure 96]

Professor Stanley McDaniel in discussion with Kynthia about her sculpture of "The Face On Mars". (Photo by Ananda Sirisena, 1994)

[Figure 97] Sculpture of "Face" made by Kynthia. (Photo by David Laverty)

THE "FACE ON MARS" - FOUR AND A HALF DECADES LATER

In 1998 JPL put out a newly obtained image of the controversial land form called the "Face on Mars" (which was originally photographed in 1976) that essentially squashed any mainstream interest in this object as a possible artificial construction. The new 1998 image, although more detailed and of higher resolution than the original, appeared to show a very ordinary plateau. In this short paper we draw attention to a newer, high resolution image (2009) whose orbital parameters were consistent with the 1976 original. We show that this new image displays details that support the earlier speculation that the object may be an intentional construction designed to depict facial features such as an eye and teeth.

It is over 40 years since the Viking orbiters photographed an unusual formation in a region of northern Mars known as Cydonia. When an image published in 1998, poorly processed by NASA, effectively gave the 'kiss-of-death' to further study of the feature known as the face on Mars, many researchers decided to leave the matter alone. However, in 2009, without publicity, an MRO (Mars Reconnaissance Orbiter) context camera (CTX) photograph was released. This is a discussion about the facial features within the CTX image of 2009.

On the 31st July 1976, the National Aeronautics and Space Administration (NASA) issued a Press Release which stated the following:

"NASA VIKING NEWS CENTER, PASADENA, CALIFORNIA
PHOTO CAPTION Viking 1-61, P-17384 (35A72)
This picture is one of many taken in the northern latitude of Mars by the Viking 1 Orbiter in search of a landing site for Viking 2.
The picture shows eroded mesa-like landforms. The huge rock formation in the center, which resembles a human head, is formed by shadows giving the illusion of eyes, nose and mouth. The feature is 1.5 kilometres (one mile) across, with the sun angle at approximately 20 degrees. The speckled appearance of the image is due to bit errors, emphasized by enlargement of the photo. The picture was taken on July 25 from a range of 1873 kilometers (1162 miles). Viking 2 will arrive in Mars Orbit next Saturday (August 7) with a landing scheduled for early September [1976]."

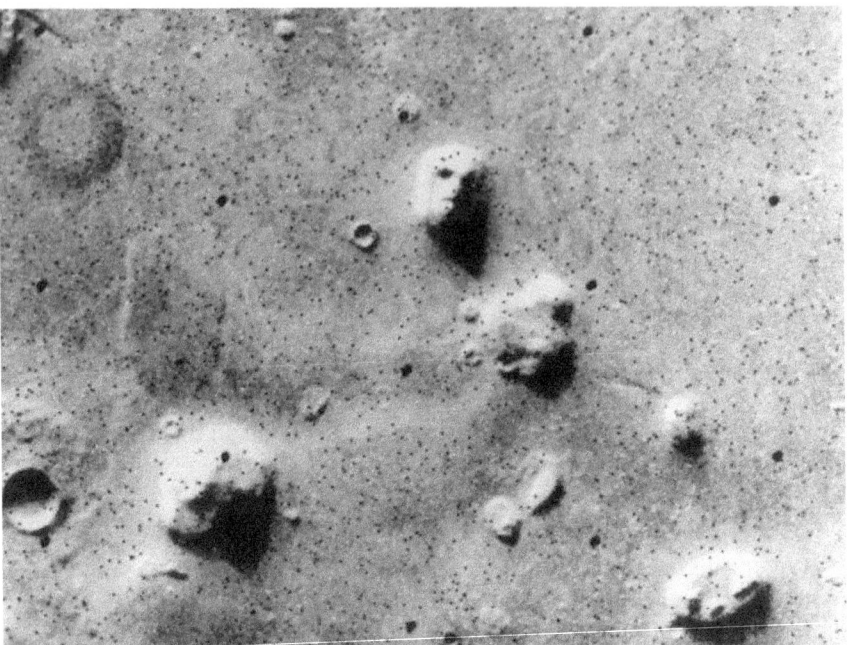

[**Figure 98**] - NASA release in 1976 of frame 35A72 - "HEAD". Taken from nearly overhead with a low Sun angle.]

The number 35A72 was a reference to Viking 1 image 35A72, the seventy-second photo taken by the 'A' camera during orbit 35. The feature, "which resembles a human head" was discovered by Dr. Tobias Owen and became famous as "The Face on Mars". It's discovery is well documented in the book, "The Search For Life In The Universe", written by Donald A. Goldsmith and Tobias Owen. (Ref. 1) In the ensuing years, the feature captioned 'human head' has received detailed study by several authors and has been referenced in various publications. (See list of references). A later image of the same feature was found on Viking 1 Frame 70A13.

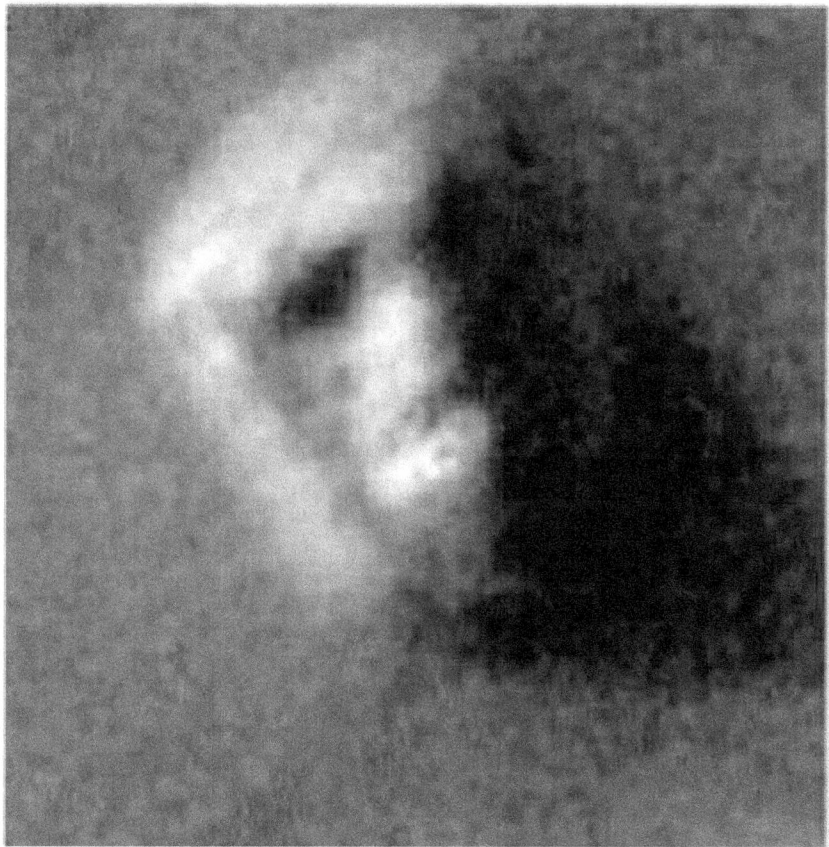

The Face on Mars from Viking frame 35A72 processed by Mark Carlotto, showing teeth in the mouth area. This was published in 1988 in Applied Optics (Ref. 7)

[Figure 99]

This author would like to thank Dr. Mark Carlotto for freely giving these enhanced images for publication. Science will progress rapidly due to the sharing of data amongst researchers and the non-biased interpretation of results obtained by space agencies.

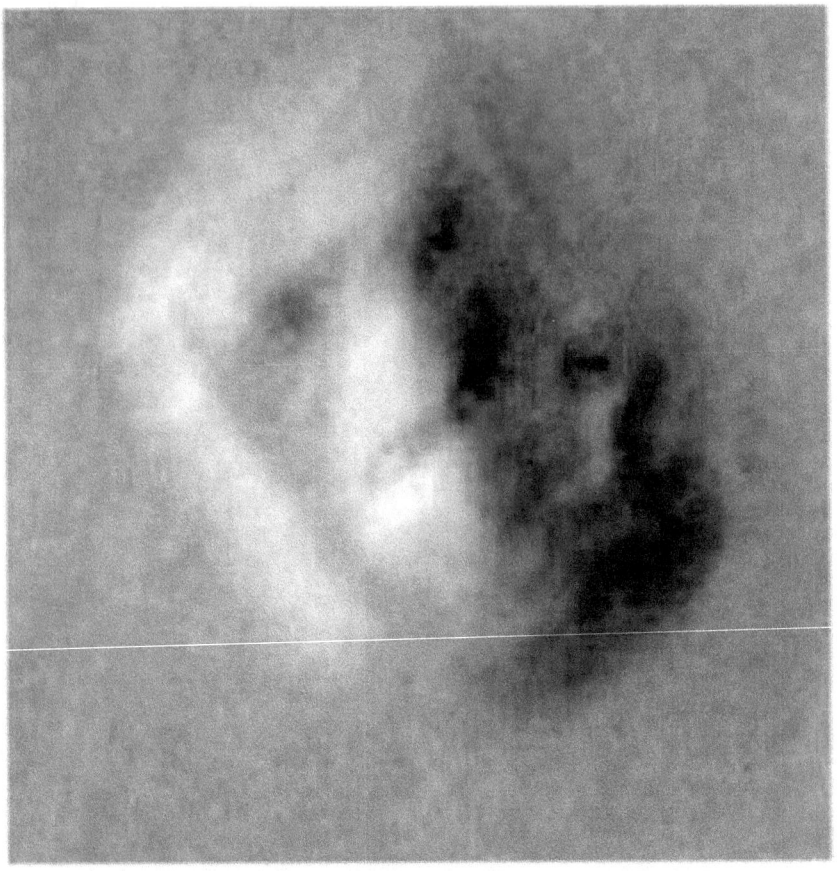

The Face from Viking frame 70A13 processed by Mark Carlotto. This was also published in 1988 in the journal Applied Optics (Ref. 7)

[Figure 100]

Event in 1995

In 1995, Malin Space Science Systems, responsible for the Mars Global Surveyor (MGS) Orbiter Camera (MOC), published an online article called "The Face on Mars", which discussed the results of image enhancements, such as bit-error correction, contrast and brightness adjustment, reseau mark removal and sharpening of digital images using bicubic interpolation, *including a cautionary tale about 'teeth in the mouth'*.

This article can be read at the following link:
http://www.msss.com/education/facepage/vikingproc.html

One section of this discussion by Dr. Michael C. Malin, Principal Investigator for MGS, tries to explain why some researchers found 'teeth' in the mouth:

"The "Face on Mars" images have been processed and reproduced many times by people with a wide range of experience and available software. However, it is important to remember that one cannot extract more information from the data than was there at the beginning.
"Two attributes that play a role in creating the appearance of "teeth" in later-processed images: first, the bit error correction "filled in" several bit errors (both above and below the "mouth") with values averaged from neighboring pixels, creating a sharper contrast in the image. Second, the jaggedness of the boundary was accentuated by this process. The lower left image (see four depictions below in Figure 2) shows how, with the application of a laplacian-sharpening filter, subtle contrast on a pixel-to-pixel level is greatly enhanced, creating a few pixels much brighter than their immediate neighbors, and much brighter than they were previously. In this image, however, the same process can be seen in many areas; the individual pixels are clearly seen and compared to other jagged features created by the pixelation of the image. Many people use interpolation when they enlarge images, however and interpolation "fuzzes" the individual pixels with their neighbours, thus "smoothing" somewhat the sharp differences between pixels. The lower right image was processed exactly the same as the lower left image, but enlarged using bicubic interpolation rather than pixel replication. *The result: the "Face" looks like it has teeth.*"

Figure shows the face from Viking image 70A13, enhanced by Malin Space Science Systems

Top left: raw image clearly showing bit errors and the dark 'reseau' mark.

Top right: cleaned-up version, smoothed to fill in the bit-errors and removal of reseau mark

Bottom left: result of pixel replication

Bottom right: result of bicubic interpolation

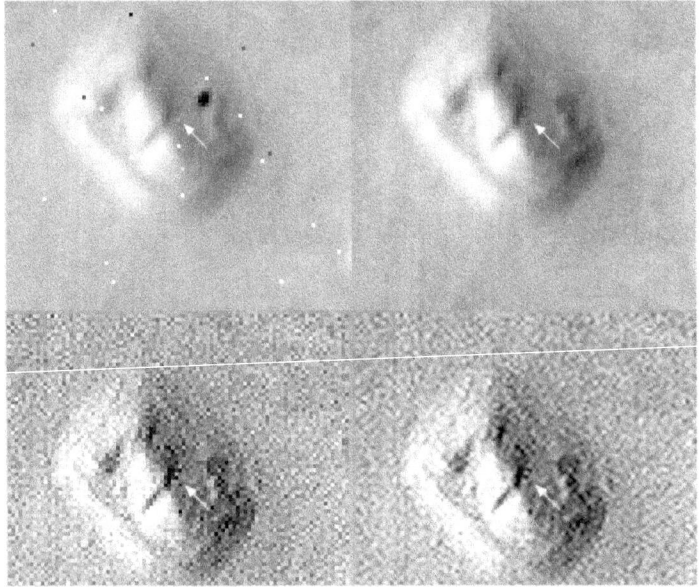

Enlarged using: Pixel replication|Enlarged using: bicubic interpolation
[Figure 101] Enhancements to Frame 70A13. Courtesy of Malin Space Science Systems, 1995]

[Notice that in the above enhancements made by Malin Space Science Systems, the arrows point to "teeth" supposedly found by bicubic interpolation of pixels in the 'left side' of the face.. These "teeth" do not coincide with those found by Dr. Mark Carlotto (and published in 1988), on the 'right side' of the feature.]

Events in 1998

On 6th April 1998, amidst a fanfare of publicity, NASA and Malin Space Science systems released an image of the face taken by the MGS Mars Orbiter Camera (MOC). The result was later posted here: http://www.msss.com/mars_images/moc/4_6_98_face_release/index.html

This image has been referred to by some as "The Catbox" because of its very weak processing. This author does not want to accuse NASA or MSSS of a deliberate obfuscation of data obtained at taxpayers expense.

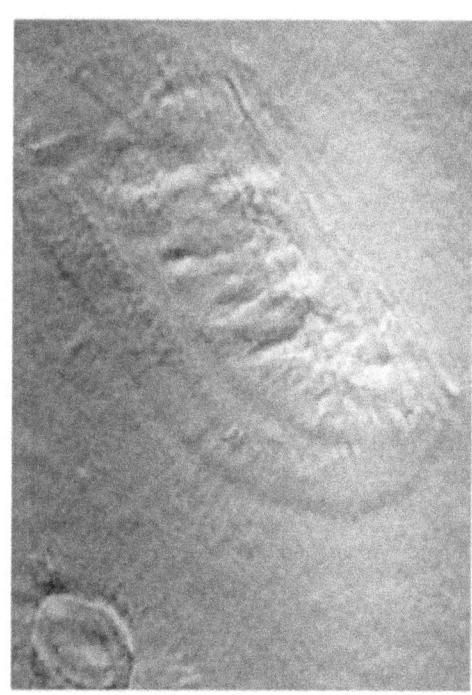

MSSS stated in 1995:
"First, given the interest in the general public about the "Face," it is appropriate to acquire such images for public relations purposes, especially since the public interest has been generated in no small way by the people who claim there is a conspiracy at NASA to withhold information from the public. Second, there are valid scientific reasons to examine landforms in the area (which, after all, is why the Viking spacecraft were photographing the area in the first place)."

MSSS stated; in 1998
"As can be seen, fortuitously, the area imaged was relatively clear, although the lack of surface definition in many nearby areas, and the low contrast of the raw MOC high resolution image, suggests haze or fog over much of the area."

[Figure 102] - NASA release of 5th April 1998 image. Poor processing made this feature look flat, featureless and without three-dimensional depth.]

Malin Space Science Systems (MSSS) stated that image processing had been applied to the image in order to improve the visibility of features. This processing included the following steps:
1) The image was processed to remove the sensitivity differences between adjacent picture elements. This removed the vertical streaking.
2) The contrast and brightness of the image was adjusted, and "filters" were applied to enhance detail at several scales.

3) The image was then geometrically warped to meet the computed position information for a mercator-type map. This corrected for the left-right flip, and the non-vertical viewing angle (about 45° from vertical), but also introduced some vertical "elongation" of the image for the same reason Greenland looks larger than Africa on a mercator map of the Earth.

4) A section of the image, containing the "Face" and a couple of nearby impact craters and hills, was "cut" out of the full image and reproduced separately, as seen below.

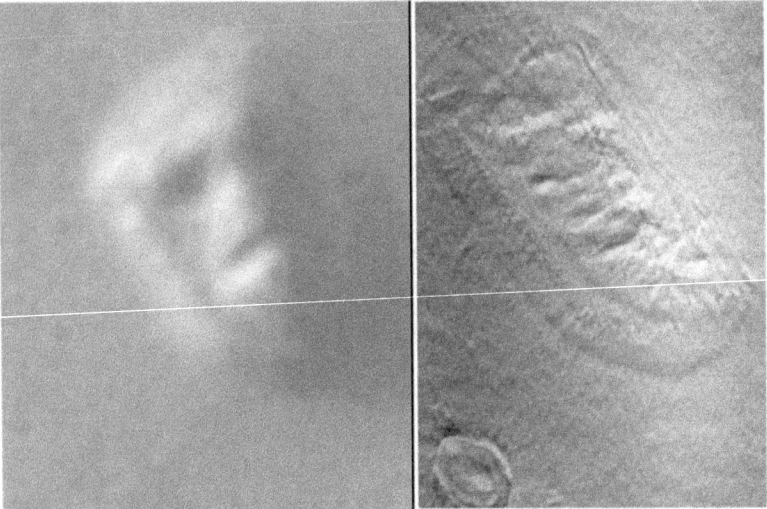

1976 Viking image at 47 m/pixel | 1998 MOC image at 4.3 m/pixel
Distance to camera - 1500 km | Distance to camera - 444 km

[Figure 103] - Viking Frame 35A72 alongside the MOC image acquired on 5th April 1998]

Notice how much more detail appears in the older, lower resolution image of 1976, compared to the 1998 release.

Image taken in 2008

Shown below, is the CTX image B01_010143_2216, acquired on 24th September 2008, from a distance of just over 311 km. This image has not been processed and is presented here as obtained from the MRO HiRise website. The incidence and emission angles are shown in the table. It shows clearly an 'eyeball' in the eye cavity and 'teeth' in the mouth area.

Center Latitude 41.68;   Center Longitude 350.77
Local time 15.31
Incidence angle 47.95; Emission angle 18.08
Image Skew Angle 90.2; Phase Angle 65.97
Slant Distance 311.29 km
Image Acquisition Time 2008-09-24 T20:34:48.242
Scaled Pixel Width 6.53 m
Image ID: B01_010143_2216_XN_41N009W

Ancillary data for CTX image acquired on 24th September 2009]

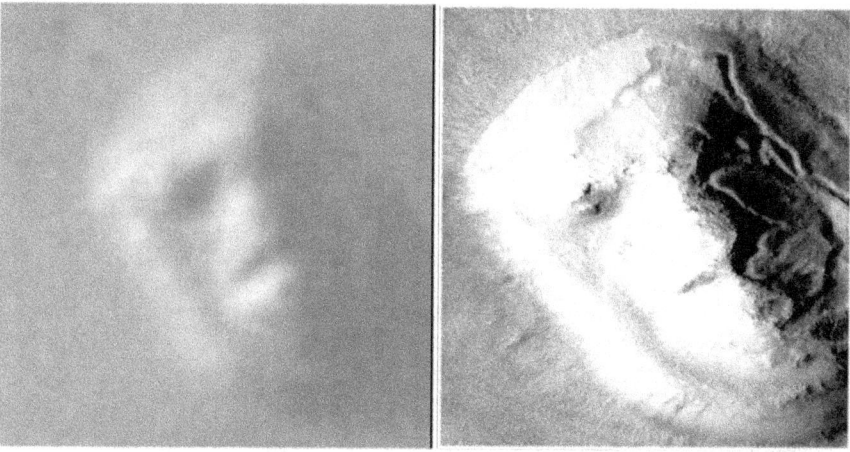

1976 Viking frame 35A72   *2008 MOC image*

[**Figure 104**] - Comparison of 1976 Vining photo with 2008 image]

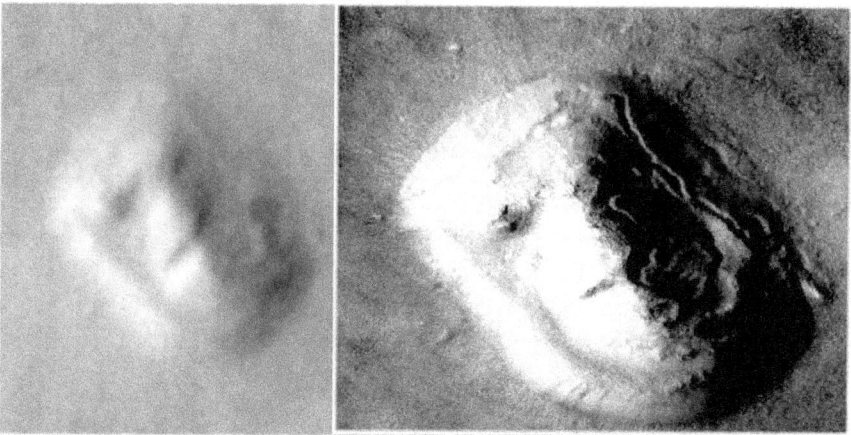

1976 Viking frame 70A13   *2008 MOC image*

[**Figure 105**] - Comparison of 1976 Viking frame 70A13 with 2008 MOC image]

Reconstruction of the Face on Mars by Mark Carlotto

A 3-D reconstruction of the feature was performed using a technique known as shape-from-shading. https://intelligentgadgets.us/docX/sfsx.shtml

A 3-D reconstruction of the face allows one to view it at different angles and orientations. Furthermore it allows one to construct time-dependent orientations so that one can imagine how the face would look to a viewer flying around the face at various positions. Shape-from-shading (also known as photoclinometry) is a method for estimating the shape of an illuminated surface from its image. For a surface of constant albedo, the brightness at a point is related to the gradients at that point by the bi-directional reflectance function. An elevation map can be estimated by integrating gradient information in the direction of the sun. The elevation map is used to re-project the original image so that it can be viewed from other look angles.

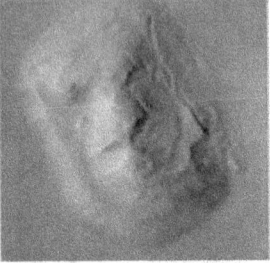 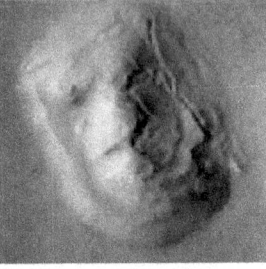 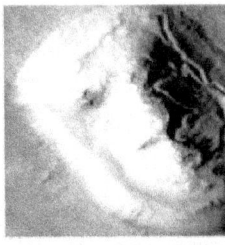

2008 image reconstituted from raw data by Dr. Mark Carlotto

Same image with contrast adjusted to highlight features of the eyeball and teeth

Comparison of same image - JPEG2000 version as previously released by Arizona State University (ASU) Mars Space Flight Facility

IMAGE ID: B01_010143_2216 ACQUIRED ON 24th SEPTEMBER 2008

Comparison of raw data with contrast adjusted image for September 2008 picture

[Figure 106]

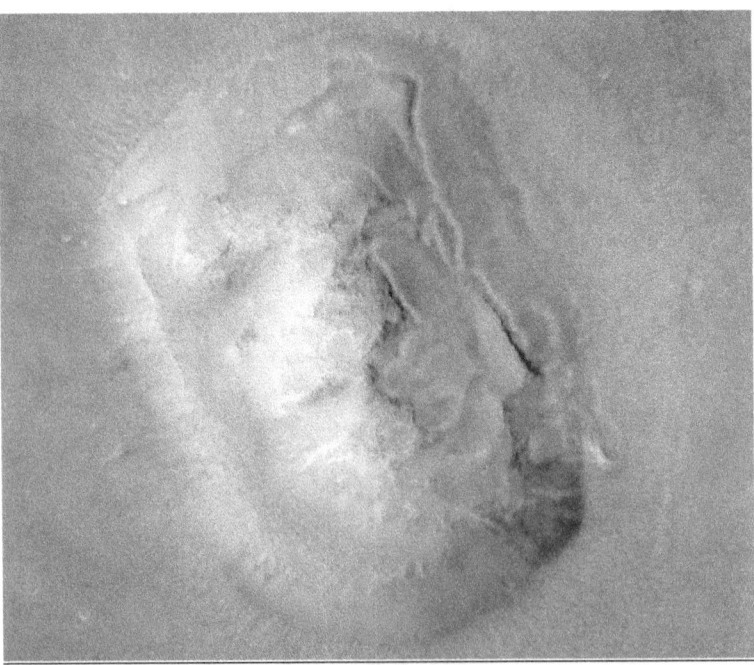

[Figure 107]

Image supplied by ESA of the face on Mars.

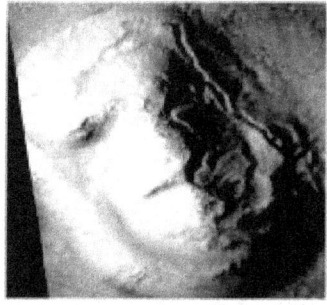
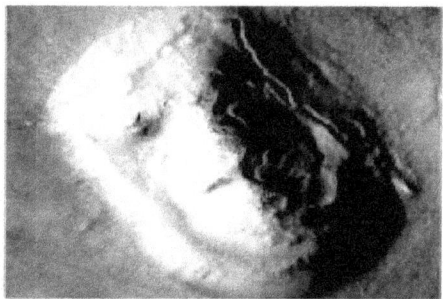

Image taken on 16 August 2008
at 15.35 local time by MRO CTX
Image No: P22_009642_2216

Image taken on 24 Septembert 2008
at 15.31 local time by MRO CTX
Image No: B01_010143_2216

BOTH IMAGES SHOW EYE DETAIL AND TEETH IN THE MOUTH AREA

Comparison of two images taken in 2008 by the MRO CTX camera.

	1976 VIKING 1 frame 35A72	1976 VIKING 2 frame 70A13	1998 MGS MOC 5 April 1998 image 22003	2008 MRO CTX frame 24 Sept. 2008 B01_010143_2216
SUNAZ degrees	294.28	277.04	Not available	174.91
INA degrees	79.93	62.59	25	47.95
S/C AZ degrees	169.05	139.54	Not available	Not given
EMA degrees	10.58	12.42	Not available	18.08
SCM m/pixel	51.73	48.13	4.3	6.53
RANGE km	1873	1725	444	311

NOTES for above Table
1) SUNAZ - Azimuth of sub-solar point. Azimuth of (direction to) North, measured clockwise from zero line defined by a horizontal line drawn from the center of the image perpendicular to the right edge.
2) INA - Incidence angle, angle between surface normal and direction to sun, measured at the center of the image. (Sun zenith angle).
3) S/C AZ -Azimuth of sub-spacecraft point. Azimuth of (direction to) North, measured clockwise from zero line defined by a horizontal line drawn from the center of the image perpendicular to the right edge
4) EMA - Emission angle, angle between surface normal and direction to spacecraft, measured at the center of the image. (Spacecraft zenith angle).
5) SCM - Image scale; resolution in metres per pixel
6) RANGE - Space craft to frame center, distance in km
7) MGS MOC - Mars Global Surveyor Mars Observer Camera
8) MRO CTX - Mars Reconnaissance Orbiter Context Camera
9) JPEG2000 - format for digital images specified by the Joint Photographic Experts Group

General discussion
Astonishing detail in the above images, including what appear to be teeth in the mouth area, suggest continued study of the "Face on Mars" is essential. If there is a chance that it represents a hominid-type face, then the implications in the SETI are profound.

All objects cast shadows. To say that shadows create an "illusion" of eyes and mouth would be illogical for this image. It is clear though, that one side of the facial feature is damaged. At 6.53 meters per pixel, this HiRise image, compared to the 47 meters per pixel of the 1976 Viking images, portrays a high-resolution photo which shows the true shape of the "Face On Mars". We see an eyeball and teeth without any bicubic interpolation enhancement.

This study is as much about facial recognition - as it might be about recognition of an alien artefact on Mars. What does one need to depict a face? An eye, a line for a nose and a mouth would be sufficient - as many advertisers for consumer products have shown. Previous findings of 'teeth' in the mouth of the face on Mars, by Carlotto and others cannot be dismissed out of hand (Ref. 7). Ananda Sirisena also found teeth in the mouth area during enhancements done to the Viking image 35A72 in the 1990's although the findings were not published at the time.

The effect of the 1998 image of the facial feature, released by MSSS, has been to draw mainstream attention away from a feature on Mars that could be of great cultural value to humanity and to our future search for extraterrestrial intelligence. SETI will have to expand away from only a radio signal detection to searching for planetary artefacts as a standard procedure. Perhaps our discovery of ETI may be closer to home than we ever thought possible.

A comprehensive epistemological study was written in 1986 by Dr. Randolfo Pozos in his book, ***"The Face On Mars - Evidence for a Lost Civilization?*** (Ref. 3). The question mark in the title of his book asked - and left it open for future research to provide an answer. Pozos states, "Is the "Face" on Mars a creation of wishful thinking or is it actually our intelligent recognition of something very important?" He reported, "One scientist had an immediate negative reaction to seeing the "Face" on Mars....nothing more should be done on this topic....since it might prove the truth of religion which was, according to his belief system, not possible."
In his book, *Planetary Mysteries* (Ref. 4), Richard Grossinger stated, about an extraterrestrial message,

"Actually a face is perfect. Compared to it, the Pioneer plaque is a lot of talk, a rather showy song and dance. Arthur C. Clarke thought of a monolith and that was a good idea. But a face is even better. A face says: if you are smart enough to find me, then you are smart enough to know that I don't belong here........There couldn't be any more clear statement of: come and see me." The idea of the Face on Mars being a calling card, or even an invitation to take a closer look at the whole region was taken up by other writers. Former astronaut and lecturer at Cornell University, Dr. Brian O'Leary, in his book *"Mars 1999"* (Ref. 5) summarised it thus, "Some scientists argue that intelligent intervention could not have done anything as bizarre as creating a humanoid face on Mars, as this would violate our operating paradigms in the SETI. I believe that such a view is overly narrow and that all reasonable inquiries into possible manifestations of extraterrestrial intelligence are worth pursuing."

Visceral reactions to the face on Mars exist even today, forty years after the discovery by Tobias Owen. When Viking project scientist Gerald Soffen had shown the print of the Face in 1976, to the press corp assembled at JPL, he had said, "Isn't it peculiar what tricks lighting and shadow can do. When we took a picture a few hours later it all went away; it was just a trick, just the way the light fell on it."

The truth of the matter was that a few hours after frame 35A72 was taken, Cydonia would have been in total darkness. Richard Hoagland recalls, in his book, *"The Monuments of Mars - A City On The Edge of Forever"* (Ref. 6), "That afternoon so many years ago at JPL, as a group of press passed around copies of the "head" photograph and laughed, someone had jokingly remarked that "the head is to tell us where to land."

Where to land on Mars? Strangely enough, one of the places initially chosen for one of the Viking landers to settle on the Martian surface was Cydonia - but it was not to be. The Viking 1 Lander touched down in western Chryse Planitia. Viking 2 landed about 200 km west of the crater Mie in Utopia Planitia. Both landers conducted three experiments each, in the search for microbial life, Pyrolitic Release (PR), Gas Exchange (GEX) and Labelled Release (LR). Each carried a Gas Chromatograph/Mass Spectrometer (GCMS). The results were surprising and interesting: Viking 1: the GCMS gave a negative result; the PR gave a negative result, the GEX gave a negative result, and the LR gave a positive result.. Viking 2: the results were both surprising and perplexing: the GCMS gave a negative result; the PR gave a positive result, the GEX gave a negative result, and the LR gave a positive result. The end results were three positive signals for life across the two sites. The GCMS has been deemed not sensitive enough to detect organic compounds; a reinterpretation of the results now suggests the samples did contain organics but the results were not understood because of the strong oxidation effects of perchlorate, a salt now known to be found in Martian soils. This fact was reported in the *Journal of Geophysical Research, 115, E12010,* in 2010.

The extraordinary image processing work done by Dr. Mark Carlotto in 1988 and published in *"Applied Optics"* (Ref. 7) showed the eye detail and teeth in the mouth as shown above. The 2008 images captured by the MRO confirm these details, predicted by Carlotto in 1988, twenty years earlier. The Journal of Scientific Exploration, in 1991, published a paper by Vincent DiPietro, Greg Molenaar and John Brandenburg espousing *The Cydonia Hypothesis.* The abstract read, "Evidence suggesting a past humanoid civilization has been found at several sites on Mars. In particular, what appear to be large carved faces, with similar details, have been found at two separate sites. Together with geochemical and geological evidence that suggests Mars was once more Earth-like in climate, the images of the objects support the Cydonia Hypothesis: that Mars once lived as the Earth now lives, and that it was once the home of an indigenous humanoid intelligence."

In 1999, four members of the Society For Planetary SETI Research (SPSR) presented finding ice in craters in Cydonia. Harry Moore, Dr. John Brandenburg, Steve Corrick & Ananda Sirisena showed photographic evidence indicating ice within craters at a meeting of the American Geophysical Meeting in the spring of 1999. (Ref. 12) NASA confirmed water ice on Mars on 22nd November 2016, seventeen years later, reported by JPL here:
https://www.jpl.nasa.gov/news/news.php?feature=6680

A book by Laurence Bergreen (2000) - *Voyage To Mars: NASA's Search For Life Beyond Earth* (Ref. 14) adopted a sceptical outlook. He stated, "The Face, *as everyone know*s, is unworthy of discussion."

That is not a correct statement. He continues, "A tenacious little community of conspiracy buffs considers this ordinary rock formation to be evidence - heck, it's *proof!* - of intelligent life on the Red Planet, the handiwork of an ancient, lost civilization, and any new data from NASA concerning the face is sure to attract their *unwelcome* attention". Despite these harsh words from Bergreen, we show here that the new data from the MRO are indeed worthy of further scrutiny. **Unwelcome or not,**

scientists do follow the evidence, wherever it leads. The authors of this paper feel that the face is worthy of continued study.

One author who has been terribly influenced by the MSSS issue of the flimsy 1998 image is Ben Bova. In his 2004 book, ***"Faint Echoes, Distant Stars - The Science And Politics Of Finding Life Beyond Earth*** (Ref. 18), he writes, "Mars Global Surveyor took up orbit around the red planet on September 12, 1997 and began mapping the surface in unprecedented detail. As far as the public is concerned, however, its biggest achievement was something of a letdown. In 1976, one of the thousands of photos that the Viking Orbiters had taken showed a rock formation that looked uncannily like a human face. Some enthusiasts, including many UFO aficionados, leaped to the conclusion that intelligent Martians had carved a monument similar to the carvings on Mount Rushmore. Skeptics wondered how Martians could carve a likeness of a *human* head, but the enthusiasts began to see not only "the face" in Viking photos but pyramids and whole cities, as well. Global Surveyors's sharper cameras showed that "the face" on Mars was actually nothing more than a heavily eroded mesa. The enthusiasts were not pleased." At least, Ben Bova does not deny that the massif does *look uncannily like a face*, unlike many others who have denied that it looks like a face and have suggested that one can "see things" that do not exist. What Bova's opinion will be to the 2008 images should prove to be rather interesting. Notice the statement, "as far as the public is concerned." Is Ben Bova not a member of the public? Is he not a member of the human race?

In 1979, three years after the Viking orbiters and landers, a daring suggestion was made in a paper by Hiromitsu Yokoo & Tairo Hoshima (1979) - ***"Is Bacteriophage ϕX174 DNA A Message From An Extraterrestrial Intelligence?"*** Icarus 38, 148-153. Here was a forward-looking attempt to consider the galactic effects of panspermia.
The extensive mathematical analyses done by Stanley McDaniel and Horace Crater concerning the mound configuration in Cydonia also need to be factored in, as the mounds are in close proximity of
the face. (Ref. 10, 11, 13, 15, 16)

Conclusion
Astonishing detail in the above image, including what appear to be teeth in the mouth area, suggest continued study of the "Face on Mars" is essential. If there is a chance that it represents a hominid-type face, then the implications in the SETI are profound. All objects cast shadows. To

say that shadows create an illusion of eyes, nose and mouth would not be correct for this image. It is clear though that one side of the facial feature is damaged. At 6.53 meters per pixel, this HiRise image, compared to the 47 meters per pixel 1976 image, we have a high-resolution photo that shows the true shape of the "Face On Mars". We see an eyeball and teeth without any bi-cubic interpolation enhancement. This study is as much about facial recognition as it might be about recognition of an alien artefact on Mars.

Previous findings of 'teeth' in the mouth of the face on Mars, by Carlotto and others cannot be dismissed out of hand. What was the motivation for NASA releasing the poorly enhanced image of the Face in 1998? Was it designed to draw mainstream attention away from a facial feature on Mars that could be of great cultural value to humanity and to our future search for extraterrestrial intelligence? SETI will have to expand from only a radio-signal search to searching for planetary artifacts. Perhaps our discovery of ETI may be closer to home than we ever thought possible.

List of references

1) Daniel A. Goldsmith & Tobias Owen. (1980) - ***The Search For Life In The Universe***
2) Vincent DiPietro and Greg Molenaar, (1982) - ***Unusual Mars Surface Features***
3) Randolfo Pozos. (1986) - ***The Face On Mars - Evidence for a Lost Civilization?***
4) Richard Grossinger, (1986) - ***Planetary Mysteries***
5) Brian O'Leary (1987) - ***Mars 1999***
6) Richard C. Hoagland, (1987) - ***The Monuments of Mars***
7) Mark Carlotto, (1988) - ***Digital Imagery Analysis Of Unusual Martian Surface Features*** - Applied Optics, Volume 27, No. 10
8) DiPietro, Molenaar, Brandenburg (1991) - ***The Cydonia Hypothesis;*** Journal of Scientific Exploration, Volume 5, No.1
9) Mark J. Carlotto. (1991) - ***The Martian Enigmas - A Closer Look***
10) Stanley V. McDaniel, (1993) - ***The McDaniel Report***
11) Stanley McDaniel & Monica Rix, Editors. (1998) - ***The Case For The Face***
12) Harry Moore, John Brandenburg, Steve Corrick & Ananda Sirisena (1999) - ***Ice Found In Craters In Cydonia.*** American Geophysical Meeting, spring 1999.
13) Mark Carlotto, Horace Crater, James Erjavec, StanleyMcDaniel (1999) - ***Response To Geomorphology Of Selected Massifs On The Plains Of Cydonia -.*** Journal of Scientific Exploration, Volume 13, No.3

14) Laurence Bergreen (2000) - *Voyage To Mars: NASA's Search For Life Beyond Earth*
15) Horace Crater, Stanley McDaniel & Ananda Sirisena (2016) *"The Mounds of Cydonia: Elegant Geology or Tetrahedral Geometry and Reactions of Pythagoras and Dirac?"* - Journal of Space Exploration, Vol 5, Issue 3
16) Mark Carlotto (2001) - *Symmetry and Geometry of the Face on Mars Revealed - "Analysis of the April 2001 Image of the Face on Mars,"* New Frontiers in Science, Vol. 1, No. 1, Fall 2001
17) Ben Bova (2004) - *Faint Echoes, Distant Stars - The Science And Politics of Finding Life Beyond Earth*. Published by William Morrow/Harper Collins.
18) Hiromitsu Yokoo & Tairo Hoshima (1979) - *"Is Bacteriophage X174 DNA A Message From An Extraterrestrial Intelligence?"* - Icarus 38, 148-153
19) Christopher Rose, Gregory Wright; (2004) *"Inscribed Matter As An Energy Efficient Means of Communication With An Extraterrestrial Civilization"* - Letters to Nature, 431, 47
20) The Society for Planetary SETI Research (SPSR), has two films (of a few seconds duration) on its website, links below, (in the Film and Video Archive) - titled : *"Face On Mars 3-D Reconstruction"* by Dr. Mark Carlotto. From MRO CTX camera raw data of
Image taken on 24 September 2008"
http://spsr.nmcc.edu/video/face.html
21*)* *"Contrast Enhanced 3-D film of the 2008 image of the Face On Mars"* rendered by Ananda Sirisena
http://spsr.nmcc.edu/video/facece.html
22) Jason Wright (2017) - *Prior Indigenous Technological Species* -, Astro-Physics, April 2017
===

This author can be contacted by email on: anandals@aol.com

Other books by Ananda L. Sirisena:

1) MASSIVE VIMANA (UFO) OVER THE ATOMIC WEAPONS ESTABLISHMENT - A Challenge For Parliament

2) OTHER WORLDS AND BUDDHISM - The Three Spheres Of Existence
(Available on Amazon books)

3) SOMETHING STRANGE ON THE LUNAR SURFACE - AN INVESTIGATION OF CRATER PARACELSUS C ON THE FARSIDE OF THE MOON

The author would like to thank all members of the
Society For Planetary SETI Research (SPSR):

Professor Stanley V. McDaniel (Founder of SPSR)
Late professor Horace Crater (who performed the initial detailed analysis explained in this book)
Dr. John Brandenburg (who took the time and trouble to travel to England to meet with my wife and myself)
Vince DiPietro, author of UNUSUAL MARS SURFACE FEATURES, (together with Greg Molenaar and John Brandenburg) who helped me obtain the raw Viking images from the Goddard Space Flight Center
Dan Drasin who introduced me to professor Stan McDaniel
Professor Jim Strange (who I had the pleasure of meeting three times)
Dr. Mark Carlotto (author of MARTIAN ENGIMAS) who I first met at a SETI conference and at the British Interplanetary Society in England
Dr. Randolfo Pozos who met with Stanley McDaniel and myself in California
Dr. David Webb who I met in Florida in 2001
Late Dr. Tom Van Flandern who I met in London when he gave a lecture at the Inner Potential Centre.
Dr. Mitchell Schwartz who provided me with valuable photos
Steve Corrick and Monica Paxson who met with me in England
Robert A. Johnston who took photos off the computer screen as I enhanced "The Face on Mars" and other images of Cydonia
Malcolm Smith who organised a conference at the British Interplanetary Society in London
Chris O'Kane who has shared a public stage with myself
David Eccott (whose opinions I value greatly)
And all the other members of SPSR, some of whom have passed on.

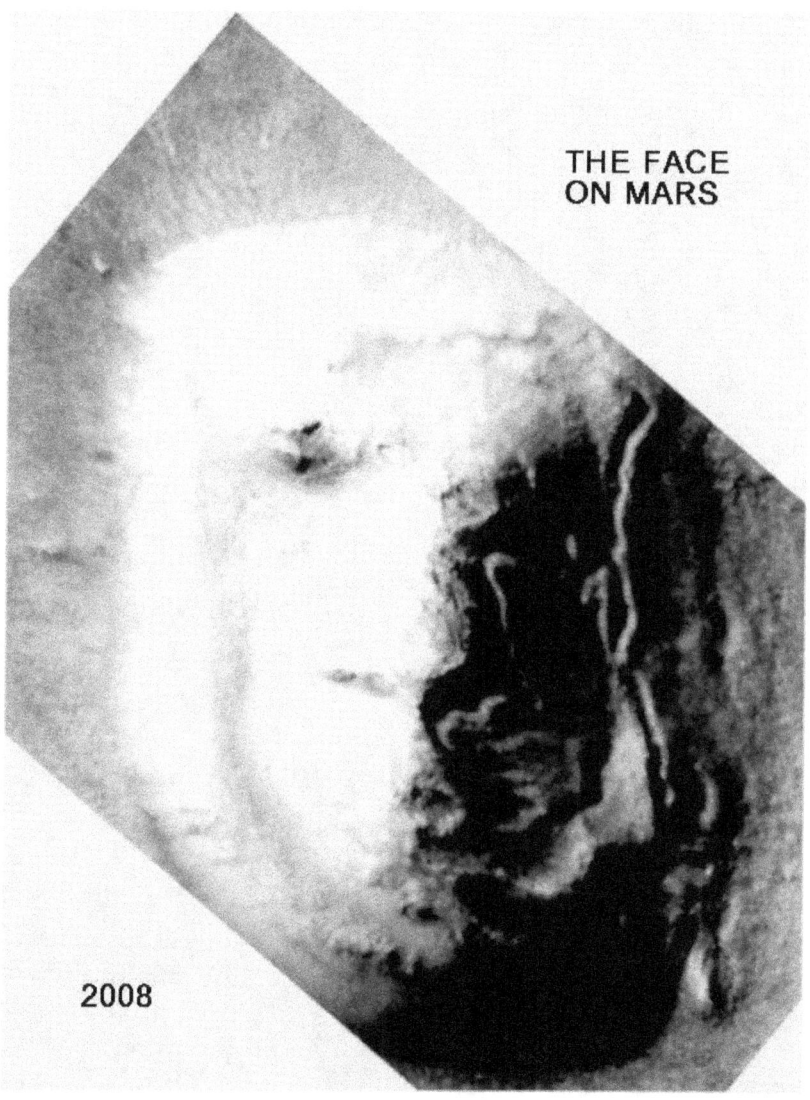

The above picture, obtained courtesy of NASA, was colourised by the author to show the details of the eye cavity and the "teeth" in the mouth of what may be a natural *massif* modified to appear like a face. Apart from rotating the image and adding colour to it, this is a 'raw' image placed on the HiRise NASA website which archives photos taken by the Mars Reconnaissance Orbiter Camera (MRO).

www.ingramcontent.com/pod-product-compliance
Lightning Source LLC
Chambersburg PA
CBHW070643220526
45466CB00001B/270